At the Forest's Edge

At the Forest's Edge

Memoir of a Physician-Naturalist

DAVID TIRRELL HELLYER

UNIVERSITY OF WASHINGTON PRESS
Seattle and London

Originally published by Pacific Search Press in 1985
First University of Washington Press
paperback edition published in 2000

Library of Congress Cataloging-in-Publication Data
Hellyer, David Tirrell, 1913–
At the forest's edge : memoir of a physician-naturalist/
by David Tirrell Hellyer.—University of Washington Press pbk. ed.
p. cm.
Originally published: Seattle, Wash. : Pacific Search Press, c1985.
ISBN 0-295-97915-1 (alk. paper)
1.Hellyer, David Tirrell, 1913–
2. Pediatricians—Washington (State)—Tacoma—Biography.
3. Naturalists—Washington (State)—Tacoma—Biography.
4. Northwest Trek Park (Wash.)—History. I. Title.
RJ43.H45 A3 2000 508'.92—dc21
[B] 99-046598

The paper used in this publication is acid-free and recycled from
10 percent post-consumer and at least 50 percent pre-consumer waste.
It meets the minimum requirements of American National Standard
for Information Sciences—Permanence of Paper for

For my wife Connie who is at the heart
of everything and always by my side.

What is man without the beasts?
If all the beasts were gone man would die from
a great loneliness of spirit, for whatever
happens to the beasts also happens to man.
Chief Seattle, 1855

Contents

Preface

SEVENTY-TWO YEARS AGO, when I was barely a year old, I showed an awareness and a fascination with living things that is characteristic of future naturalists. As I grew older my preoccupation with natural history, particularly the study of North American wildlife, increased and became my principal avocation. Throughout the years I have had frequent and unusual opportunities while hunting, fishing, camping and haunting wilderness areas to observe wildlife, often in its last refuge.

Caught in the depression and the uncertainties of the 1930s my wife Connie and I sought a way to satisfy our shared proclivities in the still largely unexploited Puget Sound country of Washington State. Perhaps atavistically searching, we found a small lake surrounded by land that was just beginning to recover from the devastation of logging followed by a great fire in 1926. We bought it in 1937 for a song. From that time on our destiny has merged with that of the lake, the rocky ridge, woods, forest edges, swamps, and meadows encompassing it. We grew and changed with it, and when the time was right, presided over its transformation into a wildlife park.

The land was raw and its future unplanned in the 1940s when I became a physician, a student of that most complex of species—man. As a pediatrician I learned to recognize the kinship of young mammals of most species. I became fascinated with the magic of mother-infant bonding, imprinting, seriousness of play, and the importance of learning and discipline for survival and fulfillment.

By the 1970s our children were grown and following their own destinies. It was at this point that Connie and I realized that our land had also come of age, and its destiny should be determined by the highest and best use for which it was suited. The concept of a protected place, where North American wildlife would find a varied and suitable habitat, where people, especially children, might experience animals in the dignity and beauty of a nearly wild state within a natural setting, began to take form. This was, of course, the dream of a child naturalist come to maturity. Our six hundred acres provided a nearly ideal microcosm for such a sanctuary. We were ready. The land was rich and waiting. The apple of insight fell on a prepared mind. The dream came true.

David Tirrell Hellyer
Eatonville, Washington
October 1985

From Admiralty Carvers to Tea Merchants

THIS BOOK IS THE ACCOUNT of my journey to Northwest Trek—of the valleys crossed, the passes topped, the side trips, and occasional times when I was lost. It deals with other times, many campsites, and special places along the way. To start I will describe some of my English roots—the Hellyer family—from which springs the rest of this memoir.

Some families seem content to remain for generations close to the land where the roots sink deep and branch widely. They have time and stability to bring forth many children who in past generations tended to follow family professions or occupations. Yet sometimes historical events, economic revolution, perhaps a new discovery, or the sudden appearance of an adventurous individual disrupt the pattern, and the members of one generation diverge to explore and experiment with new life-styles. Some such breaks in established patterns occurred in my great-grandfather's generation and have tended to persist to the present. Prior to that time, many Hellyers had been British Admiralty carvers by profession, and often several brothers were simultaneously so engaged. They did not confine themselves merely to ship's figureheads, but carved church pews and pulpits. One great-uncle carved the bishop's throne in Winchester Cathedral, while another brother carved the original figurehead of the clipper ship *Cutty Sark.* But sailing ships gave way to steam, and modern church architecture replaced the more ornate conventions of the past, and thus a traditional occupation ceased to exist, although Hellyers, as in my case, still carve occasionally as an avocation.

My great-great-grandfather had twenty-two children, all by the same woman, who, after accomplishing this almost incredible feat, died in her early forties. There must have been several sets of twins in this generation. Many of them were girls about whom little is known. Of the five boys, one was the mayor of Portsmouth, England, and three were Admiralty carvers. One departed from the pattern and signed up on the HMS *Beagle* as secretary to Captain Fitzroy, master of the ship on the famous voyage of 1831, with Charles Darwin aboard as naturalist. The *Beagle's* log, however, shows that Edward Hellyer was drowned in the Falkland Islands while duck hunting, and so his adventurous life was of short duration. Of this tragic episode Irving Stone in his fine biography of Darwin, *The Origin,* quotes him as saying, "This is a severe blow to me as it is to his messmates. Mr Hellyer was a gentlemanly, sensible young man. Now all I can do for him is write a sympathetic letter to his family in England. It's most difficult for me since I feel that the motive which urged him to strip and swim after the bird was probably his desire to get it for my collection at home.... Mr. Bynoe cut the duck from the kelp. It's a variety I've never seen before. And here is Hellyer's bullet that felled it." Surely Hellyer proved himself to be a dedicated naturalist.

My grandfather, Frederick Hellyer, was one of thirteen children. His uncle on the maternal side was a British officer by the name of Balser Alt, who, while on duty in the China Seas had contact with the Dutch merchants that had long held sole trading privileges with Japan. Balser Alt wrote his nephew, Thomas Hellyer, my great-uncle, sometime in the early 1860s, telling him of the exciting future for commercial ventures in a Japan that was just grudgingly opening its doors to the outside world. In 1859 Townsend Harris, the first American consul general and diplomat in Japan, after nearly three years was finally able to conclude a definitive treaty, which formed the basis for all other nineteenth century treaties between the Western powers and Japan. It was shortly thereafter that Thomas came to Nagasaki, where for a time he became the tutor of the Prince of Tosa's children. The prince, under the shogunate, had been one of the most powerful daimyos in the south. He governed the province of which Nagasaki was the principle city. It would have been fascinating if Tom Hellyer had recorded his experiences in the great man's household, as well as what he taught these children of a feudal age.

I remember him many years later in 1924 in London where I was recovering from rheumatic fever, but I shortly returned to Bramley Grange and never saw him again. He was slender, stooped, and distinguished looking, with a Mark Twain mustache, a black suit, and a soft black hat—a gentle and quiet man, who sat with me on a bench in Hyde Park by the Serpentine. He fed the sparrows crumbs from a hard roll he had brought along in a paper bag. We talked of eleven-year-old things. I wish today we could again share that bench and talk about grown-up things—of those early unrecorded years, and what it was really like, and especially of how he, an Englishman, found his way into the confidence of so powerful a prince at a time when Europeans were still viewed with great suspicion.

It was sometime in the late 1860s that Uncle Tom Hellyer and Balser Alt formed a trading company in Nagasaki and persuaded my grandfather, Frederick, to come out from England to join them in the venture. The company prospered and soon expanded, moving operations to Kobe as soon as that port and the adjoining area of Hyôgo were opened in 1892. Business continued to flourish. From a general trading venture it soon became involved exclusively in the profitable manufacture and export of tea.

It was at about this time that my grandfather met and married Georgianna Tirrell, the daughter of a prosperous Boston merchant, whom he met in California where they were both vacationing. He brought her to Japan, and she quickly learned to love the art and beauty so foreign yet captivating. She even made friends with some of the ladies of the Imperial Palace. In fact, she was at one point enlisted to demonstrate to them the intricacies of corsets with stays and garters—how to lace them on with the garters down instead of up. She also indicated the proper European dresses to be worn for special occasions, for it became briefly popular in those circles to abandon the beautiful kimonos for garish Victorian dress.

Since the main market for green teas, for which the

A generation of tea merchants gather around a carp pool at Miyanoshita, Japan, 1906. From left to right: Nanny, Harold Hellyer (behind tree), various cousins with their amahs, Grandmother Georgianna Tirrell Hellyer, Uncle John Liddell, Grandfather Frederick Hellyer, and Aunt Marion Liddell.

Japanese leaf was most suited, lay in Canada and the Midwest of the United States, my grandfather moved to Chicago. In 1896 he there established the head office under the name of Hellyer and Company, "renouncing all allegiance to the Queen of Great Britain and Ireland," thus becoming a United States citizen. Meanwhile, Uncle Tom was in charge of the Japanese end of the business, which was shortly transferred to Shizuoka where the factory and company houses were built. Two of their sisters, Isabelle and Minnie, married officers of the powerful and rapidly expanding Hong Kong and Shanghai Banking Corporation and settled in the Far East. After grandfather's death and Uncle Tom's retirement, my father, Harold, followed his two older brothers, Arthur and Walter, by then American citizens, and entered the family business. They were all born in Japan and their oldest sister, my Aunt Marion, was the first Caucasian child born in Nagasaki. She married a British merchant and lived in Shanghai until driven out by the Japanese invasion. She spent the remainder of her life in Hong Kong. In both cities she was the grande dame of the international community, a beautiful and elegant woman to the last.

The three generations of Hellyers who guided the company's fortunes in Japan participated in the life of that country during the time when with reluctance that isolated island nation emerged from feudalism and became the economic giant we know today. During the Second World War, the Japanese tea trade was greatly restricted, and finally in the 1960s, the company was sold. No longer will a Hellyer tea merchant or manufacturer taste the new season's leaf or compete in the marketplace for the favor of millions who find solace and civility, not tempest, in a fragrant and steaming pot of tea.

It was in April 1912 that my mother, Dorothy, father, and brother George, then six weeks old, uprooted themselves. They left Chicago and moved to Japan, where my father was to become resident manager of Hellyer and Company. For my father it was a return to a familiar atmosphere, as he had frequently accompanied my grandfather and grandmother on business trips during his youth. For my mother it was high adventure and the beginning of a lifelong love affair with the art and much of the culture of the "old Japan."

It was thus that the compound of the company's resident manager's house in Shizuoka became the site of my first investigations into the natural world. Here I discovered, as did my family, that I was indeed a naturalist, and like most naturalists, I was born with that affliction. Although frequently inherited, in my case neither my parents nor my grandparents suffered from this condition. Nor do my three daughters or my eight grandchildren seem to manifest it in a significant fashion. It is, however, important to have a parent, who, though perhaps not sharing, understands the intensity and pervasiveness of the naturalist's need to make contact with every creature that swims, crawls, flies, or simply intrudes into his environment. Many parents who have such a child do not realize that the condition is not curable, that it will not go away when he or she starts school, gets interested in sports, or falls in love. Fate or circumstance may frustrate the congenital naturalist, and he may be lulled into thinking that a cure has been effected, but if opportunity offers, the symptoms will surface as virulently as if they had never been in remission.

I consider myself a rather typical example of the congenital naturalist. My father was sympathetic, but my mother, although not herself so inclined, seemed to understand that she had a child severely afflicted with the condition, and that no discouragement was indicated. She was endlessly patient for many years. My brother George, older by eighteen months, somewhat shared my interest and was always supportive. He helped tip the balance in my favor when my demands were sometimes just beyond the borderline of reasonableness.

Naturalist Roots in Japan

I WAS BORN IN KOBE, Japan, in 1913, but we lived in Shizuoka, a moated town to the south in the midst of the tea-growing country where the family factory was located. The resident manager's house was a two-story wood frame building, western in construction, and I remember it as being ugly. It stood within a compound behind a wooden fence. At the front a wet meadow extended from the fence nearly to the front porch. There were frogs in this wet grass—so many frogs that one could scarcely take ten small steps without putting one up. They were hard to catch, but I can still feel their cool, smooth bodies and palpitating throats in my four-year-old hand and recall the ecstasy of the contact and the closeness of their bright, winking eyes.

A sandbox covered with a wide roof stood in one front corner of the compound, and an orange tree hung over it dropping small, inedible fruit in the summertime. Whenever I smell overripe oranges, I see this sandbox, the wet grass, and the bamboo hedge extending from it to the back where the goat pen and the sheds were located. It was in this bamboo hedge that my brother and I came upon a snake holding a living bird in its mouth. Armed with sticks and rocks we attacked. The snake turned, releasing the bird, and coiled, hissing. It was a long and fearful battle, but at last we emerged from the hedge, dirty and triumphant, with the snake draped over a stick. Our parents were appalled, partly I suppose by our savagery, and by fear that we might have been bitten. I remember then touching the dry, smooth, ivorylike scales and suddenly doubting the righteousness of our cause. I have killed many poisonous snakes since that time, but I have kept many more as pets and admire their singular adaptations to nearly all environments and find their varied colors and fascinating patterns beautiful.

The goat pen contained two goats that came from India. In those times there were no milk cows in Japan for the Japanese did not drink milk and found the smell of cheese abominable. We westerners obtained the goats and drank the milk, just as did my father a generation before. That, of course, did not mean that the Japanese gardener and handyman, Ojisan, or cook-san, or the amah, or Chiosan, the cook-san's wife, had to like the goats. My brother and I, however, loved them, but we treated Jane, the oldest, a large white goat, with respect because she butted. It was a condition of employment that Ojisan must milk twice daily and that one of the others should hold the goat's attention during this process by fanning her face, talking to her, and driving away the flies. One day during this perilous procedure, Jane apparently snatched the woven reed and bamboo fan from the hand of the front-end distractor and started to wave it about, causing shrieks of terror from all sides. She was considered bewitched or possessed, and no amount of reassurance would persuade the Japanese to milk again. I do not remember the outcome, but we had drunk plenty of milk by this time. Poor Jane had to put up with many indignities.

In the hot summers when the tea men were working long hours, supervising the buying, firing, hand-rolling, and drying of the tea, the oven temperatures would rise to two hundred and fifty degrees; the charcoal for drying the leaves glowed red. The fathers became exhausted and irritable, especially as they had to stop smoking to restore the keenness to nose and palate for the all-important tasting of the finished tea.

Early beginnings in Japan, 1918.
From left to right: David, Harold, and George.

Of course, I understood little of this, except it was time to take to the cool hills with my mother, older brother, and now a new younger brother, as well as an amah, J. J., and Jane to provide us with our milk. I do not remember the train trip from Shizuoka through Mishima to a junction with the steep and cobbled Tokaido road linking the ancient imperial capital Kyoto to Yeddo, now Tokyo. From that point we abandoned conventional travel to reach our rented summer cottage within easy visiting distance of Miyanoshita where the family gathered for special occasions. My recollections of this part of the journey are as vivid as the events of yesterday. We must have presented a curious sight to the Japanese, who watched our slow progress up the rough, narrow road shaded by its great cedar trees. As we gained altitude, views of the streams and gorges became increasingly wild and enchanting. Trees framed Hiroshige prints with their smooth trunks. Travel was by foot for the adults who were not too hobbled by age, long skirts, layers of petticoats, corsetry, and drawers. They walked a little, rested, and rode part of the way in swaying litters, while we children alternately walked, were carried, or rode when the adults chose to walk. Our goods were mostly loaded on packhorses. Jane's indignity reached its apogee when she was slung on one side of a horse and counterbalanced by a trunk on the other, a mode of travel which for a time must have soured both milk and disposition.

I remember those family gatherings in Miyanoshita at the Fujiya Hotel, in particular the celebration of my fifth birthday. Aunt Marion and her family were there together with the uncles and cousins, assorted nannies, and amahs. We boated on Lake Hakone and on returning to the hotel were greeted by the owner and founder of this famous resort, the extraordinary hotelier and bumptious autocrat Mr. Yamaguchi. I can still see him with his magnificent sweeping mustache rivaling in length and obviously as cherished as the six-foot-long tails of his beloved bantam roosters. The roosters sat all day on their high perches almost as though carved of wood, and I was told that at night their tails were carefully rolled in paper to keep them clean and unruffled. What Mr. Yamaguchi did with his mustache at night was beyond the scope of my childish imagination. I remember the hot springs, the carp pools, the waterwheel that turned the first generator in Japan (so the story goes), the lovely gardens, the flower palace, and the swimming

Summer retreats to Miyanoshita always included boating trips on picture-perfect Lake Hakone with Mount Fuji in the background.

pool. Yes, the swimming pool where my cousin Rosalind dropped her ring, and the pool had to be drained. I also used to climb up into the picture gallery in the tower portion of the hotel. From there I would look out over the hills, then inspect the photographs covering the walls showing Mr. Yamaguchi posed beside kings and potentates, the famous and notorious, visitors he claimed as friends although of course they meant nothing to me.

The only one of us who arrived at our destination refreshed and eager was J. J. (for Jesse Jenkins). Miss Jenkins had met the family and immediately endeared herself to us on one of our trans-Pacific voyages. She was at loose ends, having, I believe, just ended some unfortunate love affair, and was returning to Australia, where she had grown up. She was in her late thirties, perhaps early forties, athletic, forthright, and well educated. When father and mother suggested she might join our family and help with the children, she accepted forthwith, becoming one of the dearest and strongest influences of my childhood. She told us stories of growing up in the Outback, where her brothers still had a vast sheep station. She told of going in from Adelaide on camelback with Marino rams slung like Jane on their backs. These rams introduced new blood to the flocks that remained all their lives on the range, within reach of a well or billabong, for only the wool was shipped out after the shearing. She showed me pictures of the birds: cockatoos, galahs, budgerigars, finches, emus, cassowaries, and above all, the kookaburras, famed for their raucous laughter. The animals—the platypus, spiny anteater, the bandicoots, the kangaroos, wallaroos and wallabies, koalas, fierce Tasmanian devils, and tigers became as familiar to me as robins, pigeons, lions, and wolves are to most children. She talked of the "black fellows" who roamed the Outback. She made this strange, isolated land live for me. I yearned to see it.

Sadly, after three years she left us, as we were departing from Japan permanently. She felt that her life had become so entangled with ours that if she were ever to make a life of her own, she must break away. I was napping when she left—she could not bear to say good-bye. And when I was told that she had gone home to Australia, I felt utter desolation. Somehow I managed to get through the solid gate that was always supposed to remain closed, walked to the ricksha stand around the corner, and asked to be taken to Australia. I do not know what the ricksha men thought, but they knew who I was, and one of them hoisted me onto his vehicle and took me all around the moat. I must have realized the hopelessness of this quest, for when I was returned to my frantic family, I accepted the fact that J. J. was gone.

She was not really gone, for she continued to write. She described the fauna in Tasmania, where she had bought a house. She sent me books on Australian wildlife, prepared skulls of mammals, some of which I later gave to the Museum of Natural History in Santa Barbara, California. She sent me books—the last one, *The Red Center* by H. A. Finlayton, is beside me now. It is dated 1937. She must have been in her seventies at this time. An occasional shaky note arrived in the forties and early fifties and I know she died well into her nineties, still doing war with the bandicoots that devastated her beloved garden. It was partly because of J. J. that my wife Connie and I went to Australia in 1973. The spell she wove still held magic and needed exorcising or affirming. We enjoyed the fine cities, but I found all my old friends in the Sidney zoo and the game preserves. We traveled the Outback, Alice's Springs, Ayre's Rock, and it was all like a déjà vu phenomenon, familiar yet strange. With the harsh realities of the red parched earth, spinifex, towering stone shapes, and the unreality of white ghost gums, moon rocks, and emptiness, it was just as she had told me, a young but already confirmed naturalist, more than sixty years ago.

My parents had decided to leave the family business and the beloved life of Japan, because it placed such stress on my father's always fragile constitution. The Great War was not yet over. It was the spring of 1918, and my father had been unable to enlist because of his physical condition. I think this, as well as many other family factors, contributed to the decision to go to America, visit both our families in Chicago and show off our new baby brother, and then start a new life elsewhere, perhaps in Europe.

There were a number of reasons for choosing Europe and in particular Switzerland as our destination. When my father left the family business he was bought out by his older brothers, receiving, I suppose quite understandably, a lesser share. This was in the form of securities and some cash in a Chicago savings account. Although this settlement provided for only a modest income in the United States, the post-World War I dollar was worth a great deal more in Europe, and as a result, our family would be able to live much more comfortably there. They also decided to live abroad, because my mother, after her father's death and her mother's remarriage to Robert Isham Randolph, felt no close ties with home. My father, after breaking with Hellyer and Company, was also eager to try a new life in a different setting.

In retrospect, the decision was a wise one, for it gave the whole family a wonderful six years of varied experiences, which could have been equaled in no other way. The income of the securities plus the use of some capital made it possible for our family to live wonderfully well in Europe during the remainder of my father's life, while providing us a start in life on our return home to the United States.

My brothers and I were not really party to these deliberations, but were excited by the prospect of leaving Japan. The idea of new adventures, sights, places, and strange new creatures had great appeal for me, for there were few animals in our Japanese environment, and these seemed exploited. Boys captured cicadas with birdlime on long poles for no good purpose. They caught fireflies and confined them in little woven grass cages where their tiny winking lights were concentrated, but over a short time this fluorescence became dimmer and finally died out. My heart was always with the caged bear in the Shrine Garden, not the keeper; with the small, straining horses pulling impossibly heavy loads, not with the straining men and women pushing from behind just as hard for their common survival; with the stray dog who stole the chocolate cake from the dining room table, not with the father who threw a soccer ball at it in justified anger and made it yelp with fright, not pain.

All our household goods, the beautiful Japanese porcelain, furniture, and treasures, were packed with care in enormous crates. We children were vaccinated for smallpox, and I was careful to vaccinate my handsome and expensive new toy horse with its magnificent real horsehide covering. I was thorough and watched with horror as the sawdust began to pour out, and the body slowly emptied and collapsed. Whether this foreshadowed my future medical career or not, no such disaster ever resulted from subsequent human vaccinations that I have performed.

At last we were off to the railroad station in two rickshas. I had forgotten my favorite wooden doll, "Oba-San," thus compounding my general unpopularity by insisting on turning back to retrieve it—first the vaccination and then the doll! We sailed on the *Empress of Russia* and were escorted by an Allied warship on the first part of the journey, as the German raider, *Emden*, was reported to be on the prowl in that area of the Pacific. Soon we were allowed to proceed without incident to Vancouver, British Columbia. There we spent a luminous few weeks at Cowichan Lake. I saw a bald eagle's nest, heard the howls of wolves, and thrilled to the first tug of wilderness, a restlessness I did not understand. Then we went on to Chicago, where all the uncles and aunts, my grandmother, ten male cousins, and innumerable more distant connections enveloped us and tried to persuade us to put down roots among them.

I remember the false Armistice celebration and the terrible letdown, hearing of those who died during the two days before the Armistice was confirmed. Next the great influenza epidemic began. Nearly the entire population was affected. In Riverside, the beautiful little suburb of Chicago where the family lived, we and all of our cousins were sick. Two of the uncles' houses, each with a trained nurse in charge, were turned into sick bays with the children divided between. My father nearly succumbed and never really regained the precarious health he had previously struggled to maintain, and my little brother died of the overwhelming lung congestion afflicting so many of the victims of that viral strain of influenza. The rest of us survived without serious aftereffects, but my parents grieved deeply over the loss of their youngest son.

Post-war Switzerland

OUR CHOICE OF SWITZERLAND as the place to start a new life in retirement was dictated by a number of factors. First, it was one of the few European countries not devastated economically by the war; second, the climate was delightful; third, the atmosphere was cosmopolitan because of the concentration of displaced intellectuals, especially White Russians, fleeing from the chaos of that terrible self-immolation that was fought to "make the world safe for democracy." German Switzerland would not have been acceptable, as the very word "German" and the thought of the language was abhorrent in 1919, and so it was to the vicinity of Lausanne that we decided to repair. We chose a pleasant quasi-residential pension nearby called Richemont and established a base from which to explore the various life-styles available in that civilized region. There were the town villas along the shores of Lac Léman (lake of Geneva), the beautiful cities of Geneva and Lausanne, the foothills covered with vineyards, and the mountains beyond rising in tiers to the snow-capped Alps.

We met many intriguing people, mostly expatriots. I wish that I had been older and able to listen and share with more understanding in the good talks that flowed in the ambience of those postwar Swiss years when there did seem hope that men might indeed learn to live together in peace.

This was a time of extraordinary intellectual and religious ferment throughout the world. The Great War had ended, but the uncouth Bolsheviks had overthrown the archaic tyranny of the czars and were imposing a terrifying egalitarian and godless regime in its place. Refugees swarmed to Switzerland from Russia and many other troubled areas of the world. Freud had finally broken through the barriers of conventional psychology and religious beliefs. He exposed man as a vast amalgam of primitive and sexual complexes, while Adler and Jung broke away from his psychoanalytical doctrines, each preaching his own version of the role of unconscious forces in our lives. Only that happy French optimist, Dr. Coué, provided a simple, do-it-yourself approach to harnessing the forces of the unconscious. He ignored the darker undercurrents and murky subterranean depths that were the realms of his contemporaries and instead assured us that if we passed the knots of our little pieces of string between our fingers and repeated again and again, "Tous les jours de tous points de vue je vais de mieux en mieux," that all would indeed be well.

My parents were of an intellectual turn of mind, my father's perhaps more far-ranging and speculative but sometimes ignoring the inevitability that cause begets effect; whereas my mother with her extraordinarily lucid, perceptive mind, perhaps made canny by her Scottish heritage, tempered and kept the family in equipoise as various of us jumped off one side or other of the scales. My parents read voraciously and discussed new ideas and theories—everything ranging from spiritualism to new political developments—which were presented both in books and conversation.

We spent about three months at Richemont and then moved to a beautiful chalet high on the side of Les Pleiades, a rounded mountain behind Vevey, whose summit was reached by a cogwheel funicular. The two-car vehicle passed within a quarter mile of the chalet but did not stop, although it traveled at only three to four miles an hour at this point. On our way to and from the outside world to shop, my mother would hold me under one arm and jump, while my father would do the same with George. Bundles were carefully packed to withstand being thrown out ahead of the passengers.

The chalet itself was large as mountain chalets go. A covered porch supported by square posts stretched along the front and faced the descending slope. Opening onto this porch the big Dutch door and three windows let daylight into the living room. This room was heated by a beautiful tile stove and furnished with a square slate-topped table and carved chairs with backs shaped like hearts, all in peasant style. (We had left the crates from Japan in storage awaiting definitive plans.) Next to the living room was the kitchen with a deep slate sink and massive wood stove. Water flowed from a faucet fed by gravity from a neighboring spring, and baths were taken in a large tin tub in front of the wood cookstove on which the water was heated. Of course, there

Accessible only by cogwheel funicular, the high mountain chalet at Les Pleiades, Switzerland, was an idyllic haven for two young boys.

was no electricity and no inside toilet. The bedrooms were upstairs, opening onto the upper porch, which also ran the full length of the chalet and formed the overhang of the porch below. Over it all, the wide eaves brooded the structure and nearly touched the ground at the back, so nestled into the steep slope was this classic mountain chalet.

Our life on the mountain was indeed new to all of us. Already our French had become fairly fluent and although there were few neighbors to practice on, there was a small progressive school with a few pupils ranging in age from seven to thirteen and run by a strange individual named M. Nusbaum. He was a short, vigorous balding man with disturbing blue eyes that he rarely blinked. I believe he had many degrees and a deep interest in psychology, which he practiced in questionable fashion on his pupils. This we did not know at first—only that he was brilliant, egotistical, and probably charismatic to others. Frankly, he scared the hell out of me and I hid when he was about. Yet the family frequently walked the short distance to the school in the evening and shared some social activities there, and my brother was enrolled in the elementary classes.

We spent summer days tramping the hills and climbing the easier rocky spurs; we picnicked amid the narcissus that whitened the alpine meadows and learned the names of the flowers and the birds. When winter came we had luges and small wooden skis, and longer evenings to be filled with reading aloud. My brother was a natural scholar. At four-and-one-half, he was already reading the Arthurian legends and started writing in English shortly thereafter. I was not at that time a scholar. It seemed a great waste of time to learn to read because there was always someone at hand to do it for me and to answer questions at the same time. It was not for lack of curiosity but rather because reading to myself seemed too slow, as my mother and father took turns reading aloud every day for several hours. Obviously books aimed at the lowest common denominator, which was me, would have been intolerable to the others. So I was at least challenged at

Summer days were spent exploring the hills and indulging in frequent botany lessons. From left to right: mother, David, father, and George at Les Diablerets, Switzerland, 1921.

every step, and because both parents read so well I was not bored; even when I failed to understand in detail, the story in general was still clear. We read all sorts of books, all in English—fairy tales, legends, *Robin Hood, Robinson Crusoe,* the *Knights of the Round Table,* the *Swiss Family Robinson,* books about animals, all the childhood classics, and then came Kipling's *Just So Stories* and the wonderful *Jungle Books.* My brother and I could chant most of the verses and still remember them nostalgically.

It is difficult nowadays to imagine a world without radio and television, although M. Nusbaum was at that time experimenting with a small tangle of wires and a tiny crystal that emitted a crackling sound. Occasionally there emerged hints of something simulating a voice or music. This was immensely exciting to a few of the older boys, but to me the whole project seemed without promise. So entertainment was purely of the do-it-yourself variety. It is astounding how many books can be read aloud over a period of seven or eight years when at least two hours a day are devoted to it. From the time I entered the sixth grade until college, there were

Wooden skis were often the best way of getting around in Switzerland's high country.

very few books assigned that I had not read or had read to me, before. And thus the evenings were passed.

This rather idyllic interlude on the mountainside lasted nearly two years, but it became increasingly apparent to my parents that while George was forging ahead with his studies partly on his own but with some structuring from the Nusbaum school, I, now seven, showed no inclination to learn anything from books. I was quite content to listen and even more content to explore the hillsides, turning over the rocks to discover worms, insects, or sometimes a small grass-woven mouse nest. Or I would follow the shepherd about, carefully avoiding the ram, which hated everyone and respected only his master. I found streams where horsehair worms and caddis fly larvae moved their tunnel houses among the tiny pebbles, jostled by the crisp and cool water. When I tired of doing, there was looking to be done. Stretching endlessly beyond and far below the chalet lay the sloping vineyards, the settlements like beads along the lakeshore, the blue lake that stretched to the right and the left as far as one could see, showing patterns of currents and different shadings but too distant for one to see the boats constantly plying its water. Then beyond Lac Léman were the French hills and mountains rising abruptly to snow-covered heights fading like wisps of vapor in the distant haze.

If this was enough of life for me it was becoming a subject of increasing concern to others. "We must do something about David's education." And so we came down from the mountain. I do not think that it was only my state of contented ignorance that brought us down, but a general family feeling that perhaps we had isolated ourselves too much, and that it was time to become reengaged.

A Tutor's Strategy

BECAUSE OF THE difficulties of finding an appropriate school for both of us, my parents hired a tutor. It seems quite inadequate to call M. Voruz merely a tutor, for he became a close family friend to my parents and a nearly constant companion to George and me. He was a tiny man, perhaps five-feet-three or four inches tall, with dark straight hair, spectacles, a black brush mustache, and a Celluloid collar with pretied cravats. He usually wore a dark, baggy serge suit and smelled delightfully of strong cigars and himself. A bachelor who lived with his mother in a third-story apartment in Lausanne surrounded by books, he was a librarian at the university and a respected classical scholar. But he knew nothing about little boys, animals, or English. Why he accepted the challenge of teaching us I do not know. Of course, George was a delight. He read and wrote both English and French and he sat still, listened, and performed admirably. M. Voruz's problem with me was to find some point of contact that would make me want to learn to read and write. Somehow he intuitively knew that the contact could be made through a study of the environment and the living things that dwelt therein (ecology had not yet been invented). Our French was fluent by this time, and our education proceeded in that language.

We had settled in a three-story, incredibly ugly but comfortable villa called the "Villa du Bochet" facing the lake in the little town of Clarens and sited on about an acre of land. A high iron picket fence hidden in part by a laurel hedge surrounded the garden. The house was of stone and plaster probably built at about the turn of the century. We had not bought this house, fortunately, but it seemed to serve some transitional function, and we thought nothing of its past history.

It was here that M. Voruz's strategy was put into action. First we measured the entire garden and floor plan of the house, laying out the basic plan on large sheets of paper. To do this we had to add the various measurements and subtract shorter ones from longer totals. Painless and understandable arithmetic entered my ken. Next we studied what was in the garden, collected the plants, identified, drew, and colored them. In order to do this, we found out about them in books, because all this was as new to M. Voruz as it was to us. To satisfy our curiosity, we read those books, and I began to see the importance of reading. Quite suddenly the mystery vanished, the code of letters became clear. Before I knew it, I was reading and beginning to write simple descriptions under the pictures that I had so carefully drawn and watercolored in my blue cahiers. Next came the birds. They formed the subject of two more notebooks. We went on expeditions to Geneva to map and study the outflow and current of the lake where it emptied into the Rhône River. We drew diagrams of the currents and contours of the sandy ridges of the lake bottom. We took temperatures at various seasons and charted them. We kept track of rising water levels. We studied the fish and, as with the birds, drew, colored, and described them and pestered the fishermen along the shore to allow us to examine and dissect their catches. It was truly a remarkable teaching and learning experience that awakened curiosity and led our reading and investigations into many different and unforeseen byways.

For example, the year was 1975 and the month June. I stood

in front of our house at Northwest Trek Wildlife Park with Charlie Hollister, a noted geologist whose specialty is the ocean floor—its currents and composition. He is a senior scientist at Wood's Hole Oceanographic Institute in Massachusetts and should know something about the vagaries of tidal forces and other current movements in bodies of water large and small. As we watched the surface of Horseshoe Lake near our house, no wind ripples were detectable, yet several logs were floating from east to west at a steady rate of perhaps one foot per minute. On closer examination we saw that all the particulate matter to a depth of perhaps three feet (which is all that can be clearly seen) traveled with them thus proving that it was not a purely surface phenomenon. Charlie had known this lake at all seasons for nearly twenty years off and on. I asked him if he realized that every morning there is a current that travels across it from west to east, slowly, decreasing as the day progresses, and by evening reverses itself so that those logs would return to rest against the bank at the point from which they started in the small hours of the morning. Could he explain that to me? Thus challenged, he spent much of the next day confirming that this observation was indeed true. He finally said, "I remember some phenomenon in my early studies called seiche, but I don't know how to spell it. It might account for this diurnal motion." In vain we searched for seiche in several dictionaries and encyclopedias, finally consulting the eleventh edition of the Britannica and there it was, "seiche—in limnology, an irregular fluctuation of the water level of lakes, first observed and named in Switzerland."

None of these explanations seemed satisfactory, but the mention of Switzerland suddenly recalled a vivid picture of M. Voruz holding me by my belt as I leaned over the concrete breakwater in front of the Villa du Bochet, marking with a crayon the daily rise and fall of the lake level—the seiche, of course. I dug out the old notebook and here is our description as evolved by two boys, aged eight and nine-and-one-half, with a wise and wonderful little tutor who seemed ageless. "When the lake is calm, one notices small differences in level. The water falls 2 or 3 or even 20 cm., then rises and falls again with a slow and regular movement. It is caused by the fact that the air is often moister and heavier over one half of the lake and causes the water over that half to fall slowly, which in turn causes the water of the other half of the lake to rise. This phenomenon occurs on all lakes. On Lac Léman this is called a Seiche." Now I know why the logs move back and forth, and Charlie was duly impressed if not completely satisfied.

It would not be fair to imply that our entire education was devoted to a study of wildlife and natural phenomena, for we read Jules Verne, Alphonse Daudet, Dumas, and some history. There was no longer a need to motivate me for I was now launched, though far behind my brother. After classes, the evenings continued to be devoted to reading aloud in English and to the introduction of the classics, which was to continue until we attended boarding school in England. George and I also had much time to ourselves and began increasingly to speak French to each other, but always English with our parents, partly because, accentless ourselves, we found their accented French dissonant, even though they were quite fluent.

Although George and I are very different in most ways, we were close and mutually dependent for companionship. He was imaginative, quick-tempered, conscientious, and impelled to do his best in any endeavor. But he was a worrier, as are so many high achievers. While I, on the other hand, was pragmatic and not particularly conscientious, assuming I could accomplish whatever I set out to do, but not putting this to the test consistently. We fought a lot, and although I was smaller, I was quicker, and we were well enough matched so that little real damage was done. We shared more good times than bad, and when we were on expeditions together, alone, or with M. Voruz, we were eager for the other to share our enthusiasms, taking nearly equal delight in finding a hedgehog under the laurels, watching the albino blackbird that made our garden its home, or simultaneously recognizing the humor, surprise, or excitement in a book such as *Don Quixote* being read aloud to us.

As far as I can remember, our curiosity did not extend to the bronze plaque set into the stone and plaster wall at the front of the house, and I feel sure that it held only a mild historical interest to our parents when they rented the Villa du Bochet. The plaque stated that Stephanus Johannes Paulus Kruger had lived out his last years in this house and died here in 1904. "Oom" Paul Kruger, as he was affectionately known to his Transvaal Boer supporters, was a hero and patriot to the Dutch Afrikaners. On the other hand, he was an anathema to the British and in particular to Cecil Rhodes and his concept of a Cape to Cairo empire. Kruger escaped capture by the British and was taken to Utrecht on a Dutch warship. Embittered by the fact that he could not obtain support in Europe to continue his anti-British struggle, he retired to Clarens—with, rumor had it, the horde of gold he had brought out of Africa and that had never surfaced.

What follows is really my brother's story, although I shared peripherally in the profound effects. George and I slept in separate rooms, and I never saw nor heard anything unusual during the night, but some three or four months after we moved into the villa, he began to have terrifying dreams or experienced frightening occurrences night after night. As he described it, the door of his room opened, and a broad, bearded figure with wide-brimmed hat and cloak entered his room. The man stood at the foot of George's bed, while he lay scarcely breathing with horror, then slowly turned and left by the door through which he had entered. No word was uttered, no sound of steps was heard. Our parents made light of the matter at first, trying to explain away these apparitions on the basis of an overwrought preadolescent mind feeding on the Kruger legend. But the depth of my brother's anxiety and conviction was so disturbing

Disturbing nighttime visitations, real or imagined, finally drove the family out of the Villa du Bochet.

that a complete breakdown seemed imminent. Our parents sat up and watched; they also tried to reason and reassure, but without results. Finally it became obvious that we would have to move, for my brother's symptoms had progressed to severe tics, and his previously rational and stable personality had undergone a complete change.

We shall never know what really occurred during this dark and frightening period at the Villa du Bochet. Were these repeated and terrifying apparitions so vividly described and firmly believed in by my brother just hallucinations, or was a human agency at work? My brother sought evidence for the latter explanation. Some years later when he was again living in Lausanne, he found presumptive confirmation. Perhaps our innocent activities with tape measures, plans, and investigations of the house and garden had been noted and had rekindled in someone's mind a belief in the legend of Kruger gold. The night visitations may have been designed to drive us from the house so that a renewed search could be carried out. Or perhaps our moving into the house had interrupted such a search. Be that as it may, we left the villa in defeat, and it was soon rented by American tenants who in turn moved out after a very brief occupancy. Their daughter, whom I met quite by coincidence many years later, recalled that unspecified strange occurrences had also caused their departure.

Retreat to Italy

THUS DRIVEN OUT, where should we go to find the complete change of scene that we all needed so badly? A decision was quickly made. We packed only necessary clothes, leaving everything else in storage. With M. Voruz we were off for the sunny resort and fishing town of Alassio, some forty miles beyond the French-Italian border on the pleasant shore of the Ligurian Sea.

It was a new land with a new language, and the people seemed more emotional and alive than the calmer Vaudois Swiss. Here the sun shone warm and bright most of the time. When our studies were complete, we watched the fishermen bring in all their plunders of the sea. I had never seen brightly colored marine fish—marvelous sea horses, eels, and above all, the great hauls of tiny bianchetti (white bait), no larger than a kitchen match, for which I developed a great passion. It was at this time that I discovered and added the world of kitchens to my other interests. I found that in the small pensions and hotels where we stayed I was always welcome in the kitchens, perhaps because I was small, blonde, and foreign. I did not cause disturbances but always seemed to be close and curious when any dramatic culinary event was in progress. I was lifted up to watch and smell savory bubbling pots and cauldrons, allowed to pat, stroke, and dimple dough, and given special delights.

Alassio did wonders for us all. The grim experiences of

the villa seemed increasingly out of context, and George regained much of his composure and slept well again. But Alassio did not offer much historical or artistic interest. After about a month we moved on to Florence, my mother's great love. She and two friends, all called by that very popular name of the turn of the century, Dorothy, had graduated from Rosemary Hall, an excellent small boarding school for young ladies of good family, and as a finishing touch were given a "grand tour" with Miss Risser, who was the favorite teacher and classicist at the school. I have pictures of them in Cairo on camelback at the pyramids, dressed in long flowing skirts and wide hats, and in Greece, Italy, and other places. They studied art and haunted the cathedrals of Europe. They were steeped in historical England and molded and melded and exalted by the English poets. They saw richness and beauty, remembering and treasuring it all their lives, yet, like most young people of their day, who were born advantaged and protected, they were neither aware of nor really believed that their world was the unique fairyland of the very few. Out there beyond the bower and mossy wall lay a world of hunger, misery, class and cast ramparts that could not be scaled and human needs, not even acknowledged by politicians and kings.

The move to Florence, that most lovely city, was a benison for us all. We settled comfortably into a pension with a kitchen friendly to me and a fine view of the Arno River. Each day we would start out with M. Voruz and our notebooks to explore the city. We ate lunch when we were hungry and joked about the scarcity of bakeries in contrast to Switzerland where one could get a roll or brioche at almost every corner.

One cannot separate the art of Florence from its cathedrals and churches, and so we spent days on end studying the paintings, sculptures, and frescoes, and learning about the artists who bore magic names. Little boys wore hats in those days. I left a hat in nearly every church in Florence so the family finally gave up replacing them. My blonde head must have been conspicuous among the dark locks of the Tuscan children. The Duomo, Campanile, the Baptistry, the great palaces—Palazzo Vecchio, the former seat of the Medici, Palazzo Riccardi, and many others—were the background of our wanderings, while we spent many hours in the Uffizi, Pitti, and Academia galleries. Of all the Florentine artists of the Renaissance, Botticelli spoke most understandably to me. I also had a weakness for Raphael madonnas, and the magnificent colors and opulent drapings of his figures left an indelible impression on the tabula rasa of my mind. We studied Andrea del Sarto, Perugino, the Lippi, Fra Angelico, the primitives, and the sculptures, and M. Voruz's endless patience and profound knowledge made adventures of these excursions. When we tired of churches and galleries we walked the streets, examined the shops on the Ponte Vecchio, and occasionally made excursions to Fiesole on the hills behind the city, passing the Dominican convent through vineyards and olive trees to the piazza churches, villas, and gardens, all overlooking Florence, the ancient enemy below.

We began, however, to feel xenophobia in Italy. Mussolini had gained great power in 1922, and later that year became dictator after his October march on Rome. We began to see black-shirted strutting groups in the quiet streets of Florence, so once again the family decided that we should start back toward the safety and stability of Switzerland. Our final examination was to choose a favorite painting and a favorite sculpture and write an essay about them. I chose Botticelli's *Prima Vera* because its exuberant pagan beauty excited me, and George, perhaps because of our age difference, selected Botticelli's *Birth of Venus* (his essay was far better than mine). My favorite statue was that of David by Michelangelo, though I despaired of ever attaining such a magnificent body with its admirable, naked manhood from my puny nine-year-old beginnings.

It was strange that although we were nearly eight months in Italy we learned no useful Italian. Once again we studied, talked, and lived French with M. Voruz and English with our parents. All the exposure to religious themes of Italian art and the pomp and panoply of the churches apparently left no religious imprint. In our family ethics were very much a subject of discussion and a way of life, but formal religion played no part. Certainly the art made a deep impression, for I went through a period when I drew madonnas and angels and like Tennyson's Elaine "added of my wit a border fantasy of branch and flower and yellow-throated nestlings in the nest." In fact, this preoccupation lasted until our return to Switzerland when gradually madonnas gave way to animals and angels to birds as suitable subjects for my pencil.

Although we were well aware that the friendliness of the hotel personnel had changed to frank rudeness with the deteriorating political climate, we still felt that we were safe in pausing on our return to Switzerland at Bellagio for a short interval of change from the urban life of Florence. Bellagio lies on Lake Como, one of the most beautiful and poetic lakes in northern Italy. Its exotic gardens, villas, and hotels lie in the fork of the lake at the tip of a peninsula extending fifteen to twenty miles out into its deep blue waters. I remember two events that occurred in the first few hours of our stay far better than all the exursions and wondrous vistas that were the reasons for our brief visit to this enchanted resort. The first occurred as we settled into our elegant hotel rooms and opened the suitcases. Literally hundreds of cockroaches poured out, scuttling across the floor, their long antennae waving and the dry, whishing sound of the carapaces pervading the room. My mother screamed and jumped up on the bed. My father swore and pinpointed the blame on the porters at our hotel in Florence who must have devised this method of revenge on foreigners for some imagined insult or injury. George and I with the light of battle in our eyes, seized slippers and carried out one of the great campaigns in history, coordinating our attacks and uttering fierce war cries as we laid about us. It must have taken almost an hour before our last enemy lay supine and scarcely quivering. Fortunately, cockroaches are relatively dry creatures, so that the cleanup was more a matter of sweeping than swabbing.

George and I were instantly charmed by Bellagio and prepared for further adventures. That day while our mother and father were recovering from the shock, we were permitted to go outside and look around. The wide stone steps from the hotel terrace to the water were wet and slippery. I fell in. I did not know how to swim, nor did George, but since we were good friends and allies at this particular moment, especially after our cockroach victory, he quickly pulled me out and escorted me back to our rooms. It was not a good beginning. That night my father had one of his most painful and frightening attacks of enteritis.

The rest of our journey back to Lausanne was uneventful. We stayed at Richemont again while we started looking for a permanent abode nearby.

Beyond the Schoolroom

THE HOUSE THAT WE finally chose had been built by a Dutch family. It was large, square, with a flat roof and spacious rooms. Standing on about fifteen acres of land, it overlooked the suburbs behind Lausanne, the city, and the lake, and yet was only a few miles from town. The land sloped gently in terraced vegetable, berry, and orchard plantings to merge with the hayfields of our peasant neighbors to the south. They were improvident farmers, much looked down upon by our other farm neighbors, for they were constantly having babies and did not take good care of their animals or fields. My mother worried abut them and sent down rhubarb and fruit, which was supposed to improve the quality of the next baby, but other than that we had little contact with the Perrus.

Behind our house lived the Massons. They had three children—Edmond, the oldest, a strong, hardworking farm boy; then Charlie, who played the bugle and had ambitions for some urban career; and their sister, "La Mimi," a gentle, sweet girl of whose existence we were largely unconscious. In contrast to the Perrus, the Massons were thrifty, honest, and excellent farmers. Their farmhouse was a delight, with the cow sheds under the main building for winter protection and to lend warmth to the house. A single horse stall sheltered "La Lili," their gentle black mare that permitted us nearly unlimited liberties. It was a typical Swiss subsistence

Les Abeilles became home to various pets ranging from dogs and turkeys to turtles and crested newts.

farm with cows, chickens, vegetables, fruit, and enough hay, scythed by hand, to last a long winter.

When George and I were not with M. Voruz in our big, pleasant classroom, or on expeditions, we played with the Masson boys. We all got on well, and I certainly felt no class distinction as we wrestled, explored, and shared most of our outdoor activities. I can remember Charlie once asking me if we were millionaires. This distressed me, and I asked my father the question, hoping that such a shameful condition were not true because it somehow set us apart from our companions. I remember my father saying, "Well, I suppose so, in Swiss francs." (I did not report this to Charlie.) It was so soon after the war that to be American boys lent us an aura of prestige and invincibility amongst the few other local children we encountered. "Americans can do almost anything," was the mystique of those days. We did not feel particularly like American children as we had spent so little time in that country, yet there were occasions when I can remember using that cachet shamelessly to win an unfair advantage. I would pretend to be party to special knowledge, be endowed with super fighting skills, have allies that could be called on in time of need – all well-recognized characteristics of Americans.

Our house was known as Les Abeilles, the bees. Perhaps the former owners had been apiarists, but there was no evidence of this when we took possession. Certainly they had been interested in poultry, for there were extensive pens and runs to the left of the house. At the back and at the level of a large open deck that formed the roof of our schoolroom were two empty concrete reservoirs about fifteen feet on the side and six feet in depth. Not far behind these tanks was a large pine in which we soon built a fine tree house with a rope ladder. We were prepared to defend from all enemies, known and imagined, including girls (had we known any). It would be difficult to imagine a more perfect place for our family. The rooms were large and numerous with privacy for all, and my father, who was becoming increasingly frail following two bowel resections, could find quiet and solitude as he needed it.

Soon we had the poultry yard well stocked with white Wyandotte chickens. Turkeys wandered loose in the vegetable gardens and orchards, raising their own poults. The two dry reservoirs held tanks of all sizes, mostly the glass containers of old-fashioned wet batteries. These we stocked with reptiles, amphibians, and all manner of pond life collected with our nets in the many small lakes and swampy areas scattered throughout the surrounding woods and fields. We found a shop in Lausanne where we bought three turtles. I loved those turtles. We fed them tadpoles in season, and because they lived outside they thrived, avoiding the deficiencies of most household turtle pets. Since that time, turtles have often figured throughout my life, lending it their particular ageless charm, humor, and improbability. We learned the names of all the local water beetles, bugs, larvae, predators, and prey, and watched with awe the fierceness of the struggles for survival in the tiny world of weeds and water. We were given a small but excellent magnifying glass set that also opened up new vistas.

My father felt that we should learn to work with our hands. He bought a carpenter's bench and hired a local carpenter to come once a week to instruct us. The tools he brought were full-sized—heavy wooden planes, spokeshaves, augers, braces, and bits that would be museum pieces today. I had to stand on a bench to reach the work surfaces and was simply not strong enough to use the planes or effectively cut with the saws, which had little set. I hated the whole thing, but my father was a firm believer in finishing what you start, so I cried and raged my way through the building of a small portable desk with a top that lifted. Beyond doing something for my mother's birthday, when I had finished the desk, I was to be given for my very own, one of the young Wyandotte roosters. To me, this rooster seemed magnificent. When he arched his neck, flapped his wings, and crowed with the perfect three-phased challenge of the domestic cock, I felt a yearning, a passion to have him mine. I do not think I wanted to tame him or separate him from the flock: it was more a hunger of the spirit.

Well, there was a large knot at the lower edge of the pine board surface of the desk, which all of us found gave character and added charm. It seemed to me and to the carpenter to be a firm part of the wood, but it was not. Just as I had affixed the hinges and slammed the lid with finality, the knot fell out, leaving a gaping, irreparable hole. I tried putty, but that failed. I was in despair, would not consider making another top, and that was the end of the carpentry project for me. My mother tried to intervene, but this was a situation without a right conclusion. I did not get the rooster. Had I been given the rooster, I would have had no joy in him and I realized that, yet felt in some way I might have been given a chance to save face with dignity.

It is probably unfair to tell this story because it makes my family sound harsh, unfeeling, and they were not. They were devoted, intelligent parents whose whole lives were dedicated to our welfare. I tell the tale because it represents an aberration and therefore is never forgotten. As a pediatrician I gained insights from it and have often used the lesson in counseling parents about their children's innate sense of justice and explained to children their parents' unintended violations of their dignity and right to be heard.

Even if the rooster was not all mine, I continued to love the poultry. We had great success with raising hens and turkeys, learning to respect the privacy and the beak of a setting hen so as not to disturb her on the nest.

Our most exciting animal acquisition was a dog. This was the first time that we had had the opportunity to have a dog because we had been so much on the move. She was a black and white cocker spaniel puppy, and we called her Flora. Like most cockers, she was inordinately affectionate and wiggly, approaching everyone, crawling on her belly and waggling all over while leaking little streams of urine before rolling over on her back in an ecstasy of adoration. In time she became more restrained. I thought her the best dog in the world and was only once disloyal to her in my heart.

One day M. Voruz, George, Flora, and I were on our way to our favorite pond deep in the woods. We encountered there the garde champêtre (forest ranger) with his elegant little brown-white male beagle. Characteristically, Flora ran toward him, crawling and wiggling, her long black and white coat tangled and dirty. The beagle, smooth coated with symmetrical warm brown saddle and head, stood stiff-legged, tail high, the picture of dignity. For just a moment, had I been asked if I would trade dogs, I do not know to what baseness I might have descended. The moment passed. Flora rushed over to me and we embraced.

I was distracted from worrying about this episode by our arrival at the pond we had christened Lac Voruz; a deep pool about thirty feet in diameter, nearly round, and so shaded by trees and high banks on three sides that its water appeared nearly black. We assembled our nets and stood watching the dark surface, when the most beautiful newt we had ever seen slowly emerged from the depths. We had kept and studied many newts before, but had never seen a male crested newt at the height of its breeding splendor. Its sides were bright blue with spots of red, its belly orange, and along its back from head to tip of tail was a thin waving membrane that stood up displaying speckles of red, blue, and yellow. It just was not possible. We stood transfixed while a second male followed the first one to the surface, and instead of diving again they swam slowly toward the bank. With one swift scoop, they were both in the net. As the breeding season passed, our newts paled and slowly returned to the usual drabness of summer, but we did not care, for we carried with us the vision those two blue knights of the deep, sharing with us their momentary glory.

Our studies continued daily, but there was a change in emphasis. Aunt Marion, my father's oldest sister, who lived in Shanghai, had come to visit us. She was now a widow and lived with great style as the "grande dame" of the international community in her fine estate on Bubbling Well Road, where there were stables, a race track, and magnificent gardens. She had always considered herself head of the Hellyer clan since the early death of the Hellyer grandparents, and although she loved my father and mother dearly, she definitely disapproved of the informal and French-oriented direction of George's and my education.

British to the core, she was reputedly the first Caucasian baby born in Nagasaki of English parents and disapproved of her brothers having become naturalized American citizens. She started a campaign to persuade our parents to send us to an English preparatory school in Surrey. Most particularly she pointed to the severe, almost weekly "bilious attacks" that I suffered. She blamed them on our life-style, which was really a very happy one and had, I am sure, no connection with these episodes. In any case, these attacks virtually ended with puberty.

I do not mean to make Aunt Marion sound austere or dictatorial, for she was not. She had leadership qualities. We always cheerfully acknowledged her dominant status in the family and graciously enforced will. She was not particularly brilliant or intellectual, but her warmth and vitality more than compensated. She was beautiful in appearance, not tall but regal, with a clear English complexion, dark hair, and eyes of the brightest blue. Her clothes were always exquisite and her jewelry magnificent. I never saw her looking anything but perfectly turned out for all occasions. Everyone in the family loved, admired, and hoped to please her. The tone was always a little elevated when Aunt Marion came to visit the family on her endless journeys to Japan, England, Switzerland, and America.

Aunt Marion had a marvelous sense of fun and humor. Wherever she was at Christmastime, even when she was an old lady, she insisted on playing Santa Claus or Father Christmas, depending on the continent. None of the children ever penetrated her disguise. She was no ordinary Santa in a rented suit. Her red coat was velvet, her cap edged with real fur, her boots made of shiny leather, and her white beard, as fine as silk, defied detection. Her voice rumbled and bubbled with jollity.

As I think of her now, and I often do, I think first of her role as the unquestioned head and councilor of the whole family. It was nearly inevitable that the combination of her forceful personality, conviction, and persuasive arguments would tip the scales when it came to the question of our education. George was quite ready to enter fifth form, the equivalent of the seventh grade, for he wrote well in English. I, on the other hand, had never written a word of English, although in other subjects I was marginally up to second form standards. It was a terrifying wrench for us when she finally persuaded our parents that it was time for a more formal education. M. Voruz concurred that Aunt Marion was probably right. He knew that George was ready to go on in mathematics and that the companionship and competition of boys his own age was becoming important. If George alone went to school in England and I remained in Switzer-

A reunion with M. Voruz in 1965 brought back old haunts and fond memories. From left to right: David, M. Voruz, and George.

land, we would grow rapidly apart. In fact I probably needed the structured type of education more than he did. I know M. Voruz hated to let us go, but he knew that we were ready to join the educational mainstream, and that it was to be an English mainstream.

AFTER LEAVING SWITZERLAND, I did not see M. Voruz until 1965 when my wife Connie and I took a vacation trip to Europe. We stayed a few days in Brussels with George, where he was serving as a counselor to the U.S. Mission to the Common Market. Then we made a sentimental pilgrimage to Switzerland. We visited our father's grave in Lausanne and with M. Voruz, now very old, diminutive, and delighted to be with his old pupils once again, drove past Les Abeilles but did not stop. We found the chalet at Les Pleiades. We read the plaques on the wall and on the iron fence at the Villa du

Bochet where Oom Paul Kruger had lived and George had experienced his strange visitations. We laughed over old times and from time to time, M. Voruz looked at me, incredulous that I could have been the squirmy, inattentive little boy of almost forty-five years before, and now a physician, husband, and father. (George had returned to Europe frequently and maintained his close friendship with M. Voruz, while I had not seen him since I was twelve.)

M. Voruz arranged a marvelous banquet in our honor at a restaurant operated by one of his old protégés. Remembering my obsession with animals, M. Voruz arranged for a magnificent sculptured centerpiece of two giraffes fashioned out of butter. On this great occasion we sampled the best Swiss wines, which the Swiss keep for themselves—too precious to export—and topped the meal off with that wonderful nectar made by ripening a pear in a bottle, then drowning it in sweet liqueur.

George and I never saw M. Voruz again, for he died within the year. I felt deeply grateful for the chance to put an arm around that tiny man just once again, to look into his old eyes, moist with affection, that peered through incredibly dirty glasses, which he finally allowed Connie to clean for him.

A British Education

IN 1923 MY PARENTS enrolled us at Sunnydown, a good English private school and took us to London where we were outfitted using the school's list. Harrod's was the designated source for the trunk, "tuck" box, flannels (both white and gray), and round-toed sturdy black shoes with the number 55 outlined in brass nails on the soles of mine, 35 on George's. (As we knelt in chapel daily, we recognized those in front by the numbers on their shoes rather than by characteristics such as hair color, form, or posture.) We were suddenly transformed into Hellyer major and Hellyer minor.

At Waterloo Station as we started off to school, we found ourselves surrounded by groups of other boys, whose ages ranged from nine to fifteen, all dressed as we were in blue blazers with crests on breast pockets, blue cricket caps, and long or short gray flannels. My mother cried bravely; my father was moved but restrained. I remember looking out of the window as the train pulled out and seeing my retreating parents in a new light—my mother, beautiful, robust, with flaming auburn hair stood holding my father's arm. He was frail, hunched in a greatcoat that appeared too big for him. He was almost ethereally handsome and very ill. They returned immediately to Les Abeilles.

Being one of the younger boys and small for my age I was placed in the group led by a woman, Miss Eve, who acted in some ill-defined manner as a transitional figure between home and the tough, all-male school world. I did not like her or feel I needed her intervention.

Sunnydown was a typical, good preparatory school, nestled on the west slope of the Hog's Back above Guilford in Surrey and surrounded by country, woods, plantations, and deserted manor houses within running distance of the school. Along the summit of the Hog's Back ran a dirt track, part of the old Canterbury Road. I can remember seeing gypsy caravans traveling this road with their ponies, the sides of their brightly painted wagons resembling those of the western sheepherders, strung with pots and pans. The more adventurous of us learned the nooks and crannies, refuges

Sunnydown, an English preparatory school, introduced David and George to blue blazers, gray flannels, and a Spartan way of life.

and secret places of the surrounding countryside and how long it would take us to drop out of sight, run to our destination, carry out our planned projects, and return with controlled breathing and measured tread, just in time for whatever it was that we were supposed to do. For a child who had had few companions, I took to this tough all-male school world with enthusiasm.

I said that Sunnydown was a typical school, but the headmaster, Mr. Howe, though a fine teacher, was a man of terrible temper and greatly feared. He was fair, but he meted out swift punishments ranging from writing a Latin phrase a hundred times for minor peccadilloes to private or public canings for more serious infractions. To avoid such unpleasantness, the trick was to be quick, shifty, stay out of trouble when possible, plan excursions with precision, and look innocent and attentive at all times. It was a good and to me completely understandable life because I performed fairly well and had no physical abnormalities like being too fat or adenoidal. But little mercy was shown by teacher or fellow student to the boy who could not quite fit the mold: the fat boy, the stammerer, the mildly handicapped. One drum and one drummer ruled the English private school of that day.

Life was also spartan. Mr Howe did not believe in electricity. He thought it was dangerous and unpredictable, so all the illumination was by gas mantles, which were lighted with long tapers by the prefects only at dusk when the words on the page in the vaulted study hall had nearly blended with the background. Bathing was a low priority, consisting of one tepid bath a week for four boys in succession to each tubful and a cold dip daily for general morale and to encourage a clean mind. The food was frightful, but I remember being so constantly hungry that I devoured everything, and I do not remember any leftovers on neighboring plates. Even such horrors as rice and treacle, blanc mange, and suet pudding went down. This last item was the most difficult for me to enjoy for it consisted of a sagging great mound of gray, greasy dough supported by an inverted, chipped china cup. When perforated a few plums and/or figs accompanied by a gush of palely tinted fluid appeared.

Sunday breakfasts were the best meal for me, as we were each given a piece of fat, barely cooked bacon. I dreamed about that meal and would have done almost anything for a second helping—which, in fact, I did. Boxing was a basic compulsory activity three times a week, and I enjoyed it then and have as a spectator ever since. We were matched as couples for weight and skill by the boxing coach, a fine old navy fighter with a ruined face but a good way with boys. We were taught the typical European stance with that snapping straight left always out there in front, peppering and reddening the opponent's nose, while the right was cupped to cover the face and chin ready for a right cross, and the elbow protected the belly. It is still a good technique. My partner was a red-headed little boy, taller than I, equally skinny and hungry, but kinder and far less aggressive. We struck an unholy bargain, of which I am only half-ashamed even today: I could have his Sunday bacon if I promised not to hit him so hard and so often on the already reddened, slender point of his nose. We both kept to our bargain.

I loved that year and a half at Sunnydown. My English writing improved:

> Dear Mummy and Daddy
>
> I was not botom I was not 2 frome botom I was not 3 or 4 bout I beet 5 people. Mummy will you tell me wot my half te term report was like.
>
> I hope that there will be still a few triton in the lac noir you never know when chinki gets up. I hope he is all wright know.
>
> I have to go too two peoples birth day and I cant go to two peoples so I have to tchouse and it is very hard. We had 1 lovely lectur on fur traders by Captain Mansfeld oh how many hardships did that man have too indure once the cold reached 50 degris below zero

while they were thaping the mink.

We also had a rag concert on saturday oh a lovely one the masters sang and every one joined in they were all songs like it ain't going to raine no mor. and every one knows the coros and it maks a terific din.

Much love.
David.

I had nearly caught up in my other subjects. I played soccer well and was the only boy from the lower school elevated to the first team. On long runs and paper chases I never had to worry about the sting of a willow switch behind the knees as did the fat or feeble runners. There was extra time for collecting caterpillars and butterflies. Egg collecting was not then frowned upon, and I developed an extensive collection, gaining expertise with drill and copper blowpipes to empty the most fragile shell and extract the most advanced embryos through a single tiny hole. I followed the code of taking only one egg from a nest. One day on my ramblings I brought back a hedgehog and smuggled him into the basement where the bootblack, who was a friend of mine, kept him for me. I sneaked down to visit him whenever I found an empty moment, and in these ways partially satisfied my naturalist interests.

Weekdays were scheduled, but on Sundays we had more freedom. After my two pieces of fat bacon, we dressed in our heavy, gray serge suits with Eton collars and walked the three miles to Compton where we worshipped in a beautiful little Norman church under the guidance of the Reverend Gwin, who with his tall, sparse figure and snowy hair looked a credible interpreter for God. After walking home we were permitted to change and had four hours to ourselves, which to some meant books, to others sleep, but to me it meant over the hill and away through the woods to the plantations. I had no watch, but my internal clock was accurate. Then after Sunday tea we had Vespers and a short reading period before bedtime. We lived in dormitories with cots four feet apart. Under each bed there was a white potty since there were no toilets in the dormitory area. As we knelt each night with an elbow on the bed to say our prayers, we saved time by using the potty simultaneously, not sacrilegiously but because it seemed practical and perhaps more private.

It was just as life appeared to be most ordered that everything began to change. My father's health worsened. I think both he and my mother realized that he might not live much longer and that she, with her family in America, her home in Switzerland, and her children in England, would not know how to cope alone unless some plans were made. We had been returning to Les Abeilles for our vacations, which was wonderful, but the second summer vacation our parents came to England, and we spent what was to prove our last but best time together.

First we went to Wales to a lovely spot on a river where Dr. Martin, our old family doctor and friend from Japan, had a small, government-sponsored, rural practice. He took us walking in the hills with his shotgun and little spotted terrier that ignored the sharp thorns of the furze coverts and drove out rabbits, which Dr. Martin occasionally hit and gave to his patients as he made his calls. His old car would often have to stop short of the farmhouse of a patient because the track was too narrow. He was then met either by a horse and cart or a pony. I thought that the life of a country doctor was fine indeed, especially with a spotted terrier and a swift stream to fish. It was here that the cork grip of a fly rod first nestled so naturally and comfortably in my hand.

When the trip to Wales was over, and we thought nothing could match it, we went to the Norfolk Broads where a small sloop with crew was waiting for us. The crew was Mr. Applegate, who had spent his life on the water and quite characteristically could not swim. He was fine, had loose false teeth, cooked adequately, and was endlessly patient with us. Life on the Broads was delightful as we seemed to glide at ground level across the countryside through locks, sailing when there was a breeze, or gently hauling along the towpaths. We moored by the shores, took strolls to the villages to shop, and fished from the boat. Then after supper, we turned up the parrafin light and returned to reading aloud—the family delight that had been interrupted by school. We read *The Mill on the Floss,* and *Rob Roy,* as we all huddled together on the bunks in the low cabin.

When the trip on the Broads was over, and we again thought nothing could match it, we went to Salcombe on the Devonshire coast. In Salcombe we stayed at a pleasant residential hotel and had the use of a small skiff, as George and I had developed some water skills on the Broads and could both swim. The water was fairly warm. When the tide was out a great mud and sand flat extended from the seawall. Still farther out water broke over the bar that partially obstructed the harbor entrance; the same bar that was made famous by Lord Tennyson when he requested that "...there be no moaning of the bar, when I put out to sea."

One memorable day, George and I forgot to calculate the tides and were caught in the skiff aground on the stinking mud flat far from the shore. A large and very active eel, which we were neither able to unhook from our line nor subdue, shared the boat with us. The sun became more merciless; we were hungry, thirsty, and more worried as the hours passed. Ashore my mother tried to organize a rescue, but my father quite sensibly pointed out that the mud was too soft to walk on. We were quite safe if we stayed in the boat, and we would learn a lesson.

As I look back at our childhood, I realize that we were permitted great freedom to choose our adventures and were expected to find our own ways out of the difficulties into which they occasionally led us. There were some painful learning experiences, but I now heartily agree with the method and have tried to calculate and accept the possible consequences before undertaking too quixotic enterprises—but not always! In any case, time passed slowly. It was nearly six hours before the tide floated us off. We rowed back to shore, feeling put out both because no help had been forthcoming and that none had been contemplated.

All too soon the summer was over. We were back in school and our parents had returned to Les Abeilles in Switzerland. I think my father knew at this time that he did not have long to live. Decisions had to be made, and again Aunt Marion entered the picture. She wanted us to stay in England and grow up as English boys, but my mother and father were thinking of a different future. I, at least, did not realize that definite plans to move to California had been made as early as the spring of 1925, but certainly during that last summer we did know of the change. Both George and I were very sad about leaving Les Abeilles. But California sounded like a magic land, and we were naturally excited.

From my father's correspondence to her, it was obvious that Aunt Marion continued to be concerned about plans to relocate us in America. My father was sympathetic to those concerns, yet he was confident that they were not insurmountable. He agreed that life in America at that time was not entirely conducive to bringing up children, but he also felt strongly that by living a simple and reasonable life, setting a good example, and talking out problems along the way, we would learn to face the world on our own. And America was rightfully our world. We were Americans, and it was time we set down roots there. My father found for us a highly recommended, small school outside of Santa Barbara, where he hoped we would work hard and develop our own individualities. His letter to Aunt Marion was a testimony of his deep concern for our welfare.

However, there must have been further objections on Aunt Marion's part, for he wrote to her again after our summer vacation. There were to be more trips that year: one to Italy and Christmas vacation in Switzerland for one more go around at winter sports. But that need to belong somewhere was not far from his thoughts. "Personally, though I love Europe, I should be glad to go back to America and make a nest for ourselves. One really must belong somewhere, and I find that I can't belong anywhere else." He ended this second letter to her with "We are fit," as a reassurance to a concerned sister. They never went to Italy, nor did we spend our Christmas in Switzerland having a go at winter sports, for my father's enteritis suddenly became uncontrollable. He was too weak for further surgery and died shortly after writing this letter.

George and I were called to Mrs. Howe's living room at Sunnydown and told the unbelievable news. With typical English reaction to crisis, she kindly served us tea with our tears. A few days later after the funeral at Lausanne, my mother arrived in England, and we joined her at the Dysart Hotel in London. I could scarcely believe that this woman was my mother, dressed in widow's weeds with a black veil over her face, the European custom of the time. It seemed far worse than the fact of my father's death, for now we had no one. George and I stood stiffly and formally in the room. We hugged this stranger and cried a little. Then the tension broke, and I can remember that we rushed out into the hall, running and shouting, pushing and scuffling. We were horrible. We were an embarrassment, but something had to give. When we came back to the room, we were abashed and very quiet. My mother, being who she was, understood. The dam broke, and we clung to one another. From then on, individually and together, we made our adjustments and began to anticipate an exciting future while sharing a past of preciousness, warmth, and things indescribable.

George and I went back to school two days later, and our mother moved to a spacious country hotel called Bramley Grange, within ten miles of the school. She was very sad, but there were a few friends in the area, and the Howes were wonderful to her. My grandmother came from Chicago to try to lend comfort and support, but this was not really a success. She and my mother had never been very close. My poor grandmother did her best, but English hotels were cold, especially to a midwesterner, and the damp and fog were depressing. I can remember an almost endless series of coal scuttles lugged in by maids, each costing two and six, to feed the small fires in the grates. After a month my grandmother returned to the States.

My mother had only been alone for about two months when I became ill. It started with an attack of acute tonsillitis with fever. I was confined to the school infirmary, where I failed to improve. About ten days later I developed migratory and severe joint pains, which they diagnosed as "inflammatory rheumatism." I was sent home and my mother hurried me to London and the old Dysart Hotel. My Aunt Marion, who had not yet returned to Shanghai, came to the rescue. She knew both socially and professionally, an up-and-coming young Harley Street pediatrician, Dr. Patterson, whom she asked to come see me. He called, dressed in frock coat, striped trousers, and pearlgray cravat—impressive and also very competent. He confirmed that diagnosis and explained that the condition was believed to be due to a streptococcus and was now being called rheumatic fever. He further explained that there was nothing to do but rest, take large dosages of aspirin regularly, and wrap the excruciatingly tender joints with cotton wool. The aspirin helped, and even in this age of antibiotics and steroids, it is still the drug of choice. Dr. Patterson returned several times during the next six weeks and on his last visit he told me that I could go back to the country to recuperate. And so my mother and I went back to Bramley Grange.

I was a constant and demanding companion at Bramley, and my mother had little time to grieve. I insisted that she dress less mournfully and in country clothes so that we could tramp the woods and follow the hunts. There was a pack of beagles in the county that we occasionally followed, but more exciting, a fox hunt met twice a week. I had made friends with a charming, elderly widower living in the village who had been an ardent hunter but could no longer afford the upkeep of a horse. He knew the countryside intimately and the line a fox might take from almost any cover where the hounds might start it. My friend would bundle mother and me into his old car and drive to wherever the hunt met that day, take the car down the wider lanes, following the hunt. As they became too narrow we took off on foot cross-country, soon outdistancing the pony carts, bicycles, and less agile or knowledgeable unmounted followers. The hunting scene was colorful and filled with tradition. There were the horses and hounds strung out across green fields, taking fences and walls singly or in groups, the pink coats of the hunt officials, and the gatherings at the mouth of a den where a fox had gone to earth. This was always an occasion to dismount, reach for the flask and sandwiches and give the horses a breather while the small man in a cloth cap wheeled his bicycle into the center of the group, two cairnlike terriers in the handlebar baskets. Down the hole would go one of the fierce little dogs to try to drive the fox from his earth. If there was an unstopped hole at the other end, the fox would be left in peace and the pack led off to another cover to see if they could "find" again.

By cutting across the line the fox was likely to take, and perhaps by sheer instinct, we often arrived among the leaders after the kill or to the hole if the fox took to earth. My mother kept up magnificently, and I noticed that after a hunt she slept more soundly. I did not hear the sounds of grief she would try to keep from me on many other nights.

I did not like the killing and actually saw it happen only once. We had cut across a coppice and by chance were standing at the edge of the trees when we heard hounds closing fast, their voices almost blended in one frenzied cry of excitement. They were coming directly to us, but we did not see them because of a stone wall a hundred yards away that formed the boundary to the field. Suddenly over the wall came this tiny, bedraggled, and exhausted little animal, tail dragging, looking smaller than a cat. His mouth was open in a desperate snarl, and he must have known he could not reach the coppice. Close behind the sleek hound pack poured like a wave over the stones and caught him in the middle of the field as he turned to make a stand. It was over in a second, and soon the huntsman and the whip were there, the field following fast behind. I felt shaky and sick, excited, ashamed, yet proud that we were first at the kill. The huntsman blew his horn, dismounted, and with the whip, waded into the pack and raised the limp body of the fox above the reach of the leaping hounds. He looked over at me, standing a little apart, took out his knife, and cut off a front foot. He then walked over to me and drew a cross on my forehead with the bloody stump, which I was to wear until church on Sunday. I had been "blooded" and given the paw as a remembrance of the ancient initiation. My mother made me wash my face when I got home. The fox paw I kept dried and mounted on a board for many years.

It was also at this time when I was supposed to be resting and not chasing foxes that I discovered the great naturalist, Ernest Thompson Seton. Most of his books were just coming out in England in new editions. My mother ordered them all for me but doled them out so that they would last a little longer. Although written primarily for children and considered in this age too anthropomorphic to be taken seriously, they are not only fine, elegantly illustrated stories, but the life histories and the descriptions of behavior, food habits, breeding, and rearing of young are as accurate as those found in the most scientific modern studies. Anyone who has studied Seton's *Lives of American Game Animals* will encounter detailed informataion about these species found nowhere else. This fine naturalist has been nearly forgotten and deserves rediscovery. He was one of very few who talked of endangered species and the preservation of wilderness in the early 1900s.

In London and Bramley Grange, my interest in and study of American wildlife began under the tutelage of a master. Knowing that within months we would be moving to California where many of these animals I was learning to understand still lived, lent additional immediacy and relevance to this study. I read and reread all the Seton books until I knew them nearly by heart. My mother then found other books, some about African game. Relatives and friends, knowing my naturalist bent, also gave me animal books to help me pass the days of convalescence in London, and my nature library became quite impressive. But I felt I needed an encyclopedia of all the animals of the world. Finally, when my mother and I were in London, we bought the beautifully produced, large three-volume *Wild Life of the World* by Richard Lydekker, published in London by Frederick Warne and Company in 1915. Here was a feast. I started at page one of volume one and kept going, an event commemorated on the flyleaf by my notation, "London, August 12, 1926," when I was twelve years old.

The coming fall, we were to enter the Santa Barbara Boys' School. Mr. Cate, the headmaster of this school, was traveling in Europe that summer, and while in England he came to see George and me at Bramley Grange. We took him for a walk on the Downs, each carrying our Webley air pistols to shoot at the rabbits we never hit, while he described the school, the horses, and the life. He also gave us a copy of a history of America to read before we sailed. *The Battle Hymn of the Republic* was printed on the last page of this book. I thought it was the most beautiful and inspiring poem I had ever read. It helped to make me into an instant American.

We sailed from Southampton on the *Minnetonka* at the end of August. The trip was uneventful, but there was a tugging at all our hearts as the green shores of England fell away, while unknown shores lay ahead. We were met in New York by the husband of one of my English cousins.

The next morning, after traveling all night on the train, we arrived in Chicago where my mother's Uncle Will met us. Will Maclean was my Grandfather Maclean's older brother. Unlike my grandfather who had worked hard and became an important officer in Marshall Field and Company, Will was a gambler; always charming but irresponsible. During one of his gamble and lose periods, he had gone to my grandfather for a "temporary loan." This time my grandfather said no and told him to settle down in some business where he might capitalize on his wide acquaintanceship with the gambling and fast-talking segment of society. He agreed, and my grandfather set him up in a fancy tailoring business, specializing in the more gaudy but expensive clothes preferred by the gambling fraternity. Uncle Will prospered. His wide popularity was confirmed in a line in *Showboat* when one of the characters suggests to the other that he should "go to Will Maclean to get a suit made."

I am sure Will did not know how to sew on a button, but when he took one look at our English clothes—round-toed shoes, flannels with braces holding the waistline nearly up to the armpits, and cricket caps—he turned to my mother and said, "Take them to Marshall Field's and buy them some decent clothes." And that she did. For a while George and I felt very insecure with the top of our trousers at the level of our waists and only a belt for safety.

Off to California

AFTER A SHORT STOP in Chicago to visit all the relatives, we started west on the Santa Fe in luxurious accommodations. The black porters and waiters were a great novelty and were wonderful to us. It was a memorable journey through the Indian country. Towns like Albuquerque were small then and Fred Harvey stores traded Indian goods while squaws with their papooses and braves dozed on the station platforms with their backs against the walls. We stopped for water and an occasional hotbox, then the train sped on, rocketing and swaying, with the rhythmic galloping click of the rails. Black smoke belched from the stack, and the steam whistle gave forth its marvelous, mournful double call as the train raced past a whistle stop. If you happened to be standing on the small observation platform, you could hear the rise and fall of the sound as the train outran it. The cinders were everywhere. They worked their way under the closed windows and filtered past the accordion-pleated passages between the cars. They flew unerringly into eyes. All train travelers of that day were adept at rolling the corner of a clean handkerchief into an instrument to which the stinging foreign body adhered as it floated on the reddened background of a streaming eye. It was all on a grand scale.

When we reached Santa Barbara we explored the beautiful little city lying above its curving bay and sandy beaches, with mountains rising at its back, each successively more dry and rugged. From the first we were charmed. The canyon especially drew our interest. It began just above the classic Spanish mission of Santa Barbara. It curved north gently, then abruptly south, narrowing and becoming steeper until it dead-ended against the first ridge. A stream flowed down this canyon in the rainy season and held pockets of water along its course even in the driest times. The lower part of the canyon was wooded with magnificent, gnarled California live oaks, whose dark green prickly leaves gave shade and an illusion of coolness even on the warmest days. Between the oaks lay enormous rounded boulders looking like slumbering pachyderms with great wrinkled skins. It was here in the lower canyon, among these oaks and boulders, that we built our first small white stucco Spanish-style house on three acres of land. My mother was settled in a few months, and with great excitement we unpacked all the crates that had arrived from Switzerland. More by instinct than by design, the arrangement of the objets d'art assumed their same relative positions as when I first remembered them in Japan, then at Les Abeilles. Always the ancient Buddha sat in smiling contemplation on the mantelpiece (much the same smile as the Mona Lisa's) flanked by the tall blue and white Chinese vases. The brass temple candlesticks stood on the hearth, the furniture arranged itself in familiar proximities while the bureaus and tables held the same objects that had always been associated with them, and the bookshelves were full. I am sure that all who travel widely and alight briefly "nest" in this same fashion to create a sense of familiarity and home.

But while much of this was going on, George and I were off at the Santa Barbara School (now known as the Cate School), which was located about fifteen miles to the south and inland from the small ranch community of Carpinteria. The school occupied the steep flanks and top of a mesa rising three hundred feet above the fertile plain and was bounded by steep canyons to the north and south. To the east it merged with mesquite- and sagebrush-covered hills and the rugged ridges of the Coastal Range. To the west lay a wide prospect of valley, then the yellow ribbon of sandy beach white edged with Pacific surf and twenty miles of dark blue channel, which separates the mainland from the Santa Barbara Islands.

When George and I first came to the school, it had not yet been moved to the mesa top but was clustered low on the hillside facing southwest. The buildings were of simple construction–redwood vertical siding with battens and roughly finished inside. To the south, crowded along a narrow shelf above the Gobernador Canyon and creek, were the horse barns, hay storage, and corrals, while farther up the creek was the dairy where a dairyman occupied a small house and milked about ten cows supplying ample milk for the fifty boys and masters. The youngest boys like me slept on an open porch on the second floor of the longhouse. The buildings were not heated except for the fireplaces in the library and common room, so that this porch was no colder than anywhere else during the damp, foggy winter nights. The playing fields occupied the flat top of the mesa and were reached by a steep, winding dirt road, or more usually and quickly, by riding bareback up a shortcut trail.

This was the simple but very attractive school, which Mrs. Cate, the kind, brusque, Bostonian wife of the headmaster, showed us on our first day. The dry hillsides, the rough wooden buildings, the harshly accented voices of the boys, the strangeness and anxiety of the unknown, almost overwhelmed us when we were finally left to our own devices. Directly below the buildings there was a steep bank, then a green, cool-looking grove of live oak trees beside the dry creek bed. As of one mind we slid down the bank and burrowed under the soft, green bushes, which offered a more familiar, European ambience than the sagebrush hills, and began to share our doubts and fears. This helped, and when the dinner bell rang, we emerged more reconciled and joined the other boys, who welcomed us with amused but friendly curiosity.

It seemed as though we were off to a good start, but toward morning as I lay in my cot on the open porch in the midst of nine other sleeping "Jimmies" (as the youngest boys were called), I knew that something terrible was happening to me. I felt feverish, and my skin was covered with moist, itching blisters. My eyes were swollen nearly closed. I was in trouble and did not know what to do. Fortunately it was nearly 6:00 A.M., the barn bell rang, and help was at hand. The first boy who saw me gave the alarm. Mrs. Cate was called and moved me to the small section of the building that served as an infirmary. Our soft, green bushes had been poison oak. I was obviously very sensitive to this plant with which I had had no previous contact. George was totally unaffected. I spent ten days in absolute misery, then gradually the rash dried up, my eyes opened, and I was back to normal. In the years that followed, I became nearly immune to the Rhus family, and never again had more than superficial itching, even when exposed to the smoke of burning poison oak or the fine spray from a snorting horse that had been eating the plant.

All of the seven teachers at the school were outstanding individuals and two of them were particularly interested in natural history, taking pains from the start to encourage me. One was Mr. Proctor who taught French but was also a botanist. The other, Ralph Hoffman, was a well-known ornithologist, whose book on western birds is still a classic. He died climbing the cliffs on the Channel Islands in his late sixties while still birding. It was he who started me collecting and blowing eggs for the Museum of Natural History in Santa Barbara, and thus I played a small part in building up a fine collection during the next few years.

During the course of a year each of the students was expected to prepare and deliver a short speech on a subject of his choice. These talks, given in the library after prayers, were a terrifying ordeal to the neophyte. My first experience occurred only a few months after coming to school. My subject was, of course, "Birds of Santa Barbara County." I delivered my speech with quaking voice in my piping English accent from the hearth in front of the roaring fire, for it was January. The seat of my Levi's became hotter and hotter and the copper rivets on the back pockets nearly incandescent (jeans in those days always had rivets at the corners of the back pockets, as well as brass suspender buttons), so I squirmed and moved forward and back, mumbling my speech. I do not think I imparted much information, and I know my audience suffered almost as much as I.

Horses and Tea

After our first cultural shock and my first lesson in local toxic plants, George and I began to drift apart. He found a group of more scholarly friends, and although he liked camping and riding, these activities were not central to his life. Predictably, they were at the center of mine. At Santa Barbara School each boy had a horse. At 6:15 A.M. the stable bell would ring, and one had five minutes to pull on jeans, a shirt or sweater, old shoes without socks, and get down to the barn. The horses were kept in straight stalls. We talked to them and touched them as we led them to the horse trough for their morning drink and then turned them into their corral. Next we curried and brushed them until a pat on the rump left no dusty hand print. (A dark horse had disadvantages in that dusty country.)

My first horse was a fine little brown mare, well trained, and an ideal beginner's mount. Although she was only fourteen hands, she was tall for me, as I was the smallest boy who had ever attended the school. I had to stand on an apple box as I groomed her, but soon learned to vault onto her back by grasping a thick handful of mane. Riding came naturally to me, especially as we rode bareback for the short distances over and around the mesa. There is no better way to develop a natural seat than to feel the back muscles move between one's thighs, leaning instinctively with each twist and turn of the horse.

It would be difficult to overemphasize how important that horse was in our lives. We learned to take responsibility for its care, its training, to understand its limitations and its capabilities, and although maintaining control at all times, when to concede to its greater judgment on steep and difficult terrain. Some of the boys were more interested in the exacting training for gymkhanas. This took daily practice for both man and horse for the competition required skill and patience. This sport did not interest me much. To me the horse meant expanded opportunities to cover terrain, to reach distant valleys, favorite streams, special little meadows, and various campgrounds. If one did one's schoolwork satisfactorily, one could camp for two-day weekends or take day-long rides that started at dawn and ended by 9:00 P.M. We had only to sign out and specify a route and destination. In the six years I attended the school, no boy was seriously injured riding, only two or three horses died from accidents, and in that dry rocky country, where rattlesnakes were common, no boy and only one horse was ever bitten.

This speaks well for the self-reliance and judgment that this free life on horseback developed in us, when one considers the miles traveled and the days spent exploring the wild backcountry. Not only were there mountains to cross, but there was the long beach reached by riding through the walnut groves, behind windbreaks, along dirt roads to Carpinteria. There the curving beach began. One could gallop on the firm sand by the water's edge and splash through the oncoming waves. If the tide was high, one had to swim the slough, which cut across the beach halfway to the school shack where we were allowed to change our clothes.

Some horses enjoy swimming, and all of them like water, although they may hesitate to enter it if it is muddy and they cannot see the bottom. They swim with heads fully extended and bodies nearly submerged, reaching out with forelegs in a cantering motion, while hind legs scarcely kick back behind the rump. We all learned to give a swimming horse its head on a loose rein, avoiding those reaching

Each boy was responsible for his own horse at the Santa Barbara School. David on favorite charge, 1930.

forelegs, although it is safe to be towed by tail or saddle strings or to ride if you do not mind getting wet. I found riding the smooth, surging motion of a swimming horse bareback in the surf to be one of the better water sports.

I have mentioned many of the things that can be done with a horse, but not the emotional ties of affection, gratitude, and sometimes anger and frustration that form part of this strong bond. I have known many horses and quite a few mules in my time, and can think of several I truly loved, and more, that I respected for their qualities. Only a very few were ornery, infuriating, stupid, and vicious. Some experts argue that horses lack intelligence. I do not agree. Their intelligence is of a different type from that of predators such as dogs and cats. The horse is a prey animal, constantly alert to danger, surprise, or the unusual. It panics easily and is particularly upset by any entanglement of its feet, but learns readily and remembers lessons well. The mule is far more composed, will not allow itself to be overworked and takes crisis calmly. Although I have had favorite mules, they do not elicit the same affection. I know some real mule addicts, but they are not usually commercial packers who view mules less sentimentally and appreciate them more for their sure-footedness and ability to learn quickly the width of their packs.

My first two years at school were altogether satisfying. I had made a few special friends and no enemies. I had one friend, Bud Walker, who shared my interest in guns and trapping. We subscribed to the hunting, fishing, and trapping magazines. We bought traps from Sears Roebuck and sold them a few furs of raccoon, coyote, bobcat, skunk, gray fox, and nearly worthless possums that we caught. In the process we learned to recognize the tracks and other signs enabling us to interpret the activities of our foraging quarry. We learned to recognize natural crossings: the scent posts of tufts of grass that grew tall from the coyote's frequent watering, the scratch marks of hind claws. We could imagine the growl deep in the coyote's throat as he kicked the dust behind him. There were logs that were obvious highways, claw marks on trees where comfortable forks of limbs, a tree trunk, or a den made for safe daytime sleeping, and shallow pools where shiny tin on a trap pan invited the exploring forepaws of a raccoon.

In a dry country all the animals and birds, with the exception of the few rodents capable of converting their dry food into water for their metabolic needs, congregate daily at the creeks or the stagnant pools. Down the hillsides bordering these sources of water, the game trails funnel from the surrounding parched lands while the sand and mud by the water's edge reveal the kind and concentration of wild-life in an area. Therefore the Southern California ranges,

The Southern Californian ranges, mesas, and creek beds were the favorite domain of a boy on a horse.

mesas, and semidry creek beds afforded an ideal introduction to the natural history of the region while the trapping, collecting, and studying all reinforced and confirmed our own lessons. Of course, I made mistakes. Rather early on in my trapping career I caught a skunk in the figure four trap that I had set out for a ground squirrel living under the school library. We were not able to use the building for nearly ten days. The memory lingered on much longer even as the unsuccessful search for the offender gradually wound down.

One had to have special places from which to carry on one's extracurricular activities, thus shacks were an important part of our lives. There were four recognized shacks that passed from the original builders to succeeding owners by means of barter, purchase, or special arrangement. At least three more, unknown to the administration, were carefully concealed and circuitously approached where they nestled in the steep mesa hillsides. Bud Walker and I had one such shack on the north face of the mesa wall about a third of the way down the slope. This shack measured about ten feet by five feet with a dry stone wall on the bank side and rough board walls. The only opening was the doorway. A primitive double bunk occupied the outside wall, and there was a small table, bench, and shelving. On a ledge of rock to the side of the door we built a small hearth where we could cook from a tiny fire. The smoke, rising through the heavy overhanging brush, was not visible from above. Here we kept our traps, scent, and some other contraband.

Later, after my summer in Canada, I made a fine tepee, hand sewing all the eyelets, and painted it with two wide bands of red and black around the base and red animal silhouettes on the walls. I pitched this tepee far up the Lillingston Canyon on a little shaded flat beside a deep pool in a creek bed surrounded by a grove of tall, straight bay trees that made ideal tepee poles. No one, I thought, would ever discover this wonderful hideaway, but about a year after I had set it up, after scrambling down the steep side of the canyon, I found that it was gone, as were two black powder muzzle loaders, bullet molds, lead, powder horns, cooking equipment—everything. I never found a trace of any of my gear, nor did hints or rumors ever surface. I have since had tepees made on the same pattern, but never had the energy to hand sew another one.

School was not all extracurricular activities, however. Scholastic standards were high and expectations were usually met, for to fall below them implied severe curtailment of privileges. Latin, the classics, and one or more foreign languages were taught. Great emphasis was wisely placed on history. All of us were headed for college, Harvard being Mr. Cate's school of choice, but Yale, or even "lesser institutions" like Stanford or Berkeley, were acceptable. This changed in later years, when eastern school applicants were in the minority.

All these activities, these new discoveries, and this entirely new and fascinating western life, had driven most of my nostalgia for Switzerland and England from my consciousness. I was rapidly and deliberately losing my English accent.

Meanwhile, at home in Mission Canyon, many changes had also occurred. My mother met and was attracted to a completely charming native Vermonter, who had migrated to Tacoma, Washington, in 1889, where he joined and eventually became president of the Wheeler-Osgood Company, a large plywood and door manufacturing firm. He had retired at the age of sixty-two, after his wife's death, and moved to Santa Barbara in a restless search for a new life. His name was Thomas Emerson Ripley. Accounts of his youth and early years in Tacoma starting in 1889 are delightfully recorded in his two books: *A Vermont Boyhood* (D. Appleton-Century Co., New York, 1937) and *Green Timber* (American West Publishing Co., Palo Alto, Calif., 1968). Tom Ripley was completely captivated by the quiet, lonely, and very beautiful thirty-nine-year-old widow with the copper-colored hair. And although there was an age difference of twenty-three years, they did not look mismatched.

I think that Tom was one of the most handsome and most distinguished men I have ever known. Even at ninety, he remained impeccable in dress, close-shaven, his white moustache precisely trimmed. He never lost his humor, his talent for storytelling, nor his nearly perfect gift of recall.

In *A Vermont Boyhood,* he described himself as graduated from Yale, class of '88, having completed a four-year course in Glee Club, yodeling, guitar playing, tap dancing, and other branches of the fine arts. What he did not mention in his book, however, was that his preoccupation with the more frivolous side of school life at Andover had failed to gain him admission to Yale. When he heard that he had not been accepted he did not dare tell his father, a Civil War general of intimidating rectitude. Therefore, Tom packed up like his classmates and reported to Yale as though he were regularly enrolled. His hand was up to answer all the questions. He volunteered, he submitted extra work, and it was not until two weeks later that a professor noticed that the name Thomas Emerson Ripley did not appear on the rolls of the class of '88. He was told to report to the dean to whom, using his native sincerity and charm, he explained his predicament. So eloquently did he plead his cause that he was granted a further two weeks to make up his deficiencies. His family and friends never knew the truth about his unorthodox admission to Yale University.

After his retirement from Wheeler-Osgood Company he added archery, bow making, the creation of beautiful furniture, sketching, and painting to his other accomplishments. Above all, he had a way with people that drew to him everyone he met, regardless of age or station in life. If he saw someone across the room whose appearance or actions interested him, he would simply walk over, introduce himself, and soon another friendship was cemented. Occasionally throughout the years I said to myself, "This time you're going to be rebuffed." But I never saw it happen. Tom was the antithesis of my father—gay and lighthearted, he loved

Dorothy Hellyer with her second husband,
Tom Ripley in later years.

to be with people, dance, sing, and play the guitar and the flute. My father, on the other hand, was serious, solitary, and reflective by nature, cut off from more strenuous physical activities by a frail and unreliable constitution. He and my mother had been deeply in love and shared a depth of feeling and community of interest, but had little opportunity to establish long-lasting friendships and few occasions to enjoy moments of frivolity.

It was with no sense of being untrue to a first love that she and Tom became engaged. George and I succumbed rapidly to Tom's charms and welcomed the obvious new happiness that swept my mother along. Soon they were married, with us as their sole attendants. This formality completed, the four of us had a good lunch together. We then went on to the important task of buying a new headstall for my horse, truly a "bridle" party we all agreed.

Tom and I were great friends. He may have been a strict and rigid father with his own children, but with me he never played the heavy father role. I talked to both my mother and Tom freely with little reserve, and they never hesitated to give guidance and advice from their differing experiences. Although I was away from them much of the time at school or on some summer excursion, when we did come together, particularly at teatime with their friends or mine, we gathered about a big stone fireplace in the huge room Tom had added to my mother's house. Usually tea was served at the round table on the stone-flagged patio in front of the big living room window that looked out toward the woods. There was always room for one more and gradually a widening circle of friends began to gather. Eventually a group formed of all ages, who rarely needed to visit one another, for it was easier to say, "I'll see you for tea at the Ripleys."

No one can calculate the gallons, the lakes of good China tea that my mother poured during her lifetime. Yet those endless cups drunk in different places and the civilized atmosphere that surrounded that gentle ritual were always conducive to good conversation and manners, a three-generational activity that enriched us all, spanning the gaps so often considered unbridgeable. I rarely missed tea when I was home. Characteristically, my mother thought that people came to tea because of Tom's magnetism, and he as firmly believed my mother's charm the centripetal force.

It was at tea that I first got to know Belmore Browne, the fine painter of mountain landscapes and North American animals; Donald Culross Peattie, the author-naturalist who lived just across the road; and Richard Bond, the encyclopedic scientist, whose knowledge of natural history was profound, and who tutored me in mathematics and became a lifelong friend. These last two naturalists wrote letters that helped gain me admission to the University of Chicago Medical School years later.

Tom Ripley knew England almost as well as my mother. He had once been the foreign representative of Wheeler-Osgood and had soon become a member of Brooks Club and a friend of many of the influential and attractive people on the current London scene. He had also traveled on the Continent, so it was only natural that he and my mother were eager to return for a leisurely three months of travel, revisiting and seeing favorite haunts through each other's eyes. It was then the third year of our new life in America. George had planned activities for that summer, whereas I, at fifteen, seemed too young to be left at loose ends. So a wonderful plan was evolved. I was to spend the entire three months with Belmore Browne in the Canadian Rocky Mountains where each summer he sketched and painted. We would then return to Santa Barbara where, as director of the art school, he could finish his summer fieldwork in the more leisurely atmosphere of the studio.

Belmore Browne and the Canadian Rockies

I ARRIVED IN BANFF with a new Woods down sleeping bag, an inflated estimate of my horsemanship and campcraft, and boundless anticipation. We were to spend the first two weeks getting the horses shod and patched up after their long winter spent on the range in the Panther Valley. They wintered there well, and the mares were bred by good stallions that were turned out to run with them by the Canadian government. They often had a foal at their side come spring. The front hooves of some of these colts turned inward almost at right angles from following their mothers in deep snow. As they stepped in the more widely spaced maternal tracks, their feet tended to slide together to accommodate their narrower chests, and in the process their small hooves inverted. The feet of these colts straightened out in a few months when the outside edges of the hooves were severely trimmed, and they were allowed to follow their mothers on the summer trails. Minor cuts and bruises from fighting or accidental falls or scrapes were also doctored.

Ten days before we were to start on our trip, I decided that the nice quiet little pale roan horse Belmore had found for me was much too tame and dull a ride, so without asking permission I rode out to the Stony Indian encampment in the wide grass and scrub willow flat below Mount Rundel. This Indian gathering occurred yearly in early summer when the tribe would come in from its nomadic travels and set up a fascinating tepee village, full of noise, activity, and color. There were many ponies, some loose, some hobbled with thongs, and some picketed. Small Indian boys riding bareback dashed between the individual family lodges, dogs barked, and smoke from many small fires rose in fine, bluish wisps. The smells were a pungent mixture of slightly tainted meat, humanity, horse, and excrement with a strong ammoniacal pervasion.

With rising excitement, I sat my horse at the edge of the meadow. This was Indian life as it must have been for centuries. Most of the bucks and squaws wore fringed and beaded buckskins as at least part of their clothing. The tepees, although made of canvas rather than hides, were for the most part uniform in size, approximately twelve to fourteen feet in diameter and the same in height, for a tepee is formed from a half-circle of material with the wind flaps added. The few much larger tepees presumably served for ceremonial functions. Many were painted with broad stripes around the base, and a few were decorated with symbols or designs. All of them displayed the fine shading of smoke stain graduating from nearly white at the base to the dark brownish black of the smoke hole at the peak. Usually ten or more poles fourteen to sixteen feet in length projected above the canvas, making up the ribs that supported the walls. Two longer and slimmer poles fitted into the pockets of the elegant smoke flaps that lend such style to a properly proportioned tepee. The purpose of this Indian gathering was social, but it was also commercial, for it was the only opportunity for them to obtain goods such as buttons, thread, guns, ammunition, traps, and wonderful sticky things to eat. They in turn brought furs, beaded goods, buckskins, and horses to trade.

As I watched, I hoped that I could acquire a buckskin shirt as well as a horse more to my liking. Accordingly, I looked at all the horses grazing or being ridden around and through the camp. Finally, my eye fell on a tall, snorty, slightly Roman-nosed buckskin gelding. He was a rich dun color with a fine black stripe down his back and a black mane and tail. A young Indian was riding him bareback. As he moved, he elegantly plucked up his big splayed feet, which had never been trimmed or shod. When he broke into a canter he was full of energy. Had I been more experienced, I would have realized that he was barely green broke.

My mind was made up. I rode into the encampment and up to the rider of my new horse, and with a lot of hand motions and "how muches" and denying that I wanted to trade the roan I was riding, he understood and said, "Twenty dollars." Then I pointed to his filthy but admirable fringed buckskin shirt and went through the same charade, offering my Levi jacket to sweeten the bargain—"seven dollars." I had just enough money left over from my train trip north so the deal was consummated. I unsaddled Roany and gently placed blanket and saddle on "Buck's" back. He shivered and humped a little, and I felt my first qualms, but he stood quietly while I tightened the cinch. As I was about to change bridles, the Indian shook his head and insisted that I keep the old headstall and bit that he was wearing. I agreed, mounted without incident, and leading Roany, started to ride home, mostly by cross-country trail.

When I arrived, Belmore was working on some chore by the corrals. The expression on his face should have warned me that all was not well. He asked, too quietly, what I intended to do with this new horse. "Ride it on our trip," I said. "Didn't you like the other horse I got especially for you and that was well accustomed to the rest of the string?" There was a long silence, then he walked over, looked at Buck's slightly cracked but sound hooves, noticed the high head carriage, the rather wide-eyed look, and flared nostrils, and said, "Take him into the corral, unbridle and unsaddle him, and put this halter on."

I did as I was told, not appreciating how far out of line I had been. As I started unsaddling, I realized that the buckskin jacket, at nose level where I had tied it behind my saddle, smelled very unpleasant. On looking more closely, a large number of body lice were busy escaping from its folds where free meals were no longer being provided. I hung it over the corral rail to be dealt with later, unsaddled without incident, and then reached out to slide the headstall over Buck's ears—and it happened. He ducked his head, swung it sideways, nearly knocking me over, then went straight up, pawing and rearing until the rope by which I had fortunately tied him to a corral post brought him down. I was completely unprepared, but had enough horse sense to know that he was wildly head shy. I started again, working my hand slowly up his neck and reassuring him in unaccustomed English that I had meant him no harm. All went well until I got close to his ears, and up he went again. I have always been patient with animals, and I could hardly turn to Belmore for help, so again and again I tried my cautious approach with no success. Eventually I severed the cheek straps with my knife, just above the bit, so it would fall off.

Later I learned that the Indian squaws usually catch and bridle the horses, and some of them have no patience with a head-shy animal. To save time, they approach such a horse, hand over hand along the picket rope, whack it hard behind the ears with a concealed picket pin, and cause the animal to drop its head in a momentary daze. This does not improve the situation for future owners unaccustomed to such techniques.

I did not think I would have better luck with the sturdy gray leather halter, but to my surprise, Buck did not object to my slipping it over his nose, passing the head strap over his neck, and buckling it as far below his ears as possible. Then by working the cheek straps on each side, everything slid into place without incident and the rest was easy. I found two harness snaps and fastened them to the bit on either side. He accepted the bit without trouble, so I was able to attach it, reins and all, to the cheek rings of the halter. Never again, so far as I know, was that halter removed, and the head shy problem was thus solved. Although I had not been aware of it, both Belmore and his wonderful wife, Agnes, who treated me so lovingly and well, had been watching from the window. I am sure she interceded on my behalf, for when I walked into the kitchen there was a large glass of milk and a fat slice of chocolate cake waiting.

The buckskin shirt—a mix of moosehide, caribou, deer-skin, and green and red beads—was taken to an unappreciative cleaner and returned smelling, as it still does today, of woodsmoke and ill-defined human tanning products—not unpleasant—and I wore it in the mountains all that summer and many, many times since.

Preparation for packtrips into rough wilderness country, especially those of many weeks' duration, called for careful planning. Sawback packsaddles had to be repaired, while latigos and other gear also required maintenance. In this region, all the saddles, both pack and riding, were

David on horseback at Belmore Browne's corrals in Banff, Alberta, Canada.

center-fire or Montana style single rig, with soft mohair or soft cotton cinches. Breast straps and other such gear were dispensed with despite the roughness of the country.

Provisioning in those days was relatively simple, for there were no dehydrated or freeze-dried meals. Staples were flour, oatmeal, sugar, baking powder, salt, onions, potatoes, carrots, rice, dried fruit and a few hard apples, bacon, klim (a powdered whole milk), jerked meat, jam, syrup, and canned butter, all pretty heavy by modern standards. Belmore used apple boxes for packing, slinging one on each side of a horse with a loop of light rope. A pack cover protected the load, and a one-man diamond hitch finished the job. (When packing horses or mules, it is always wise to be sure the kitchen gear, horse bells, and so forth, do not rattle, for they may spook a horse, certainly annoy the riders, and spoil the quiet of the wilderness experience.)

So we started out, six packhorses with White Rabbit, a placid albino gelding, carrying Belmore's awkward stretched canvasses and paintboxes, Belmore on his wise little mare, Kelpie, and I on my high-stepping buckskin horse. It did not take long in 1928 to escape all traces of civilization in the Alberta Rocky Mountains. By midafternoon we were far into a valley with swampy meadowland on one side and a rough, forested ridge on the other. Timberline was perhaps five hundred feet above us on the sketchy trail we were following. There had been an intermittent drizzle most of the day, but the sun had just broken through. I was riding at the rear of the packtrain. Ahead the horses had stirred up a yellow jackets' nest in an old log across the trail, but as often happens in cool, wet weather, it takes time for the colony to assess the danger and retaliate. The warming sun, however, brought them out in an angry, swirling formation, concentrated in a patch of sunshine. Into this I rode unknowingly.

The next moment, everything came apart. My horse must have been stung many times. He started a bucking, plunging run, and my saddle turned under his belly. I peeled off in an arc that fortunately threw me clear of flying hooves.

By this time, Buck, with saddle under his belly, was off and down the trail in the direction from which we had come. Whether it was the speed of my departure, the odor of my shirt, or the wasps' furious pursuit of the horse that saved me, I do not know, but I was not stung.

Belmore heard the commotion from the head of the line. He called to me to follow on foot until we came to a wide spot in the valley where there was some horse feed and a little stream. It was a poor camping site, but evening was approaching, and we had to attempt to retrieve my horse and saddle (although I am sure that Belmore would have been glad to let the horse go). Silently, he indicated where we should camp and left me to unpack the horses, get out the hobbles and bells, then wait miserably in the drizzle and increasing gloom, for I had not yet learned my duties nor the routines that were expected. I did gather wood, covering it up with a pack cover, built a small stone fireplace, and started a fire. I felt guilty and alone in this unfamiliar, northern mountain country of wolves and bears. I feared that in the dark Belmore might not be able to find his way back to me, and I would be left hungry and solitary to face the terrors of the night.

About two hours later, one of the packhorses raised its head from grazing, pointed its ears down the trail, and whinnied. There was an answer, and in a few minutes, Belmore and Kelpie, with Buck now wearing the saddle in the appropriate position, came to the edge of the firelight and I went to help. This was a turning point. I told Belmore how sorry I was for all the trouble I had caused, told him that I wanted to learn and help, and not go my own way and show how much I knew. He was wonderful. It was too late to set up a tepee, but we cooked up something, crawled under pack covers, and slept. In the morning the sun was out and life began again. Belmore told me that he had found Buck standing in the middle of a swamp with the saddle still miraculously under his belly. He had allowed himself to be caught from horseback and seemed to welcome the resaddling.

All the second day we rode through a widening valley, traveling just at the timberline. There were bighorn sheep, mostly ewes with a few new lambs on some of the higher slopes, and at that distance they did not seem afraid, but turned to watch us, then bunched, moved restlessly, and looked again. We saw mountain goats twice that day, but on sighting us they climbed on and upward along nearly invisible ledges and fissures with the deliberate steady pace that is characteristic of this species.

That night we camped early by a wide stretch in the stream. There had been a camp there before us as tepee poles were carefully stacked across logs to keep them from rotting, making it a simple chore to set up our own tepee. Yes, just right for a twelve-footer. I think the standard size of tepees and poles must be one of the earliest examples of the technology that developed from the surely serendipitous discovery of the benefits of interchangeable parts. As we unpacked the boxes and belled three of the horses, they each whickered at the sound of the bell they recognized as their own. Belmore hobbled four, picketed one, and the rest were free. They settled contentedly to feed on the abundant grass in the meadow beside the stream, after first having a good roll in a bare patch of earth, where a shallow pool had stood earlier in the year.

I had not made a very good start, but I was determined to be helpful and learn my duties. Belmore quietly outlined them in order. First, spread out the sweaty horse blankets to air and dry. Then while he was setting up the first three poles for the tepee, gather wood, starting with dry twigs of pine and spruce with the bright orange-brown needles adhering. This he called "Indian kerosene," and as any woodsman knows, it flares into crackling flame at the touch of a match. Next, larger twigs; then cooking wood one to three inches in diameter and about a foot long, which was easily obtained by breaking off the dead lower limbs of trees, or from the forest floor; and finally, a few larger chunks for an after dinner fire. (An ax is rarely needed to gather wood for an overnight camp in northern country. Most of the wood suitable for cooking can be broken over a knee or struck against a rock or the trunk of a tree to reduce it to proper size.)

After I had gathered the first twigs and kindling, I stopped to watch Belmore raise the tepee. First he set up a tripod of three poles lashed together loosely enough so that they could be spread apart to make the beginning of the twelve-foot circle. He laid more poles between the forks at the level of the lashing, equally spaced except where the door was to face—away from the prevailing breeze, so as to carry away smoke from both the outside cooking fire and from the fire in the tepee. (The smoke flaps could be adjusted to accommodate a shifting breeze.) Next Belmore attached the reinforced tab at the center of the semicircle of the tepee to a pole and carefully raised it to rest in a space saved for it directly opposite the door and stretched the canvas around the poles to the doorway. There he fastened the two halves together both above and below the round doorway by interweaving six-inch green sticks through and back out of the overlapping eyelets. The tepee took form, and all that was then left to do was to spread the poles to make the walls tight and symmetrical, attach the flap door, push the longer slanting poles into the pockets at the end of the smoke flaps, then peg the bottom of the tepee to the ground. From that time on I needed no teaching. At each new camp Belmore would place the tripod. We would each set poles, spread canvas, and in about ten minutes the elegant, altogether harmonious shelter that is the Indian tepee would be ready to occupy.

Inside we placed a small log across the back some distance from the canvas wall and made a bed of boughs, convex side uppermost, covering them with a pack canvas. We put our sleeping bags in this fine bed, then brought in our personal gear, stacking it against the wall. A tiny stone fireplace beside which small wood was piled for slow feeding into the fire, for lighting a pipe, and in my case a Bull Durham cigarette, completed the furnishings. All leather goods were hung off the ground to keep them from porcupines, and any food attractive to bears was either brought into the tepee or slung from a branch. We always built a stone cooking fireplace after skinning off the grass or forest duff and only cooked in the tepee if the weather was stormy.

Meals were simple and followed a pattern long established by Belmore's lifelong mountain experience. Lunch on the trail usually consisted of a sandwich of two thick buttered pancakes with slices of barely cooked bacon and big white onions as filler. It sounds awful, but to a growing boy on the trail it was altogether satisfying. In camp this was accompanied by hot, sweet tea, while on the trail cool stream or lake water was enough. Every three days Belmore made a bannock of oatmeal and flour with a pinch of soda or baking powder, salt, and some bacon drippings. We cooked this in a reflector oven in front of the fire making a delicious bread about an inch and a quarter thick in the eight- by twelve-inch pan.

For supper fish was our mainstay supplemented by bacon and the rather small amount of dried meat and beans we carried. Therefore it was my pleasurable chore to keep us supplied with trout. After supper I washed up the dishes at the creek or lakeshore, the oatmeal pot was set on the back of the fire, covered, and there it simmered and bubbled until the fire died out. In the morning the thick crust was stirred in over the breakfast fire with the addition of a little water. Powdered whole milk was reconstituted; sugar and fat bacon completed the meal. This oatmeal bore little resemblance to the anemic "quick" oats of the supermarket. About every third day we varied this fare with pancakes, syrup, and usually trout.

By the third day we were deep in the mountains, and I began to feel a wonderful sense of having become part of an entirely new experience, more than just an extension of a short California camping trip. There was an exhilaration, a feeling of heightened awareness. The little lakes we passed were either emerald with darker streaks where shadows of

David sporting the Indian buckskin shirt and a fish caught from Sawback Lake in the Canadian Rockies.

high clouds passed over, or a clear dark blue if pocketed in a hollow where the sun did not reach. Always there were the mountains with their cliffs, screes, and immense boulders piled one on another. Patches of snow and ice fields separated the greening high sheep meadows. Streams rushed, turbulent over rocky steep stretches, flowed smoothly and deep between cut banks where flat meadows of dwarf willows, grasses, and sedges lay on either hand, or bubbled and sparkled in the wide stretches where sand and pebble bars divided courses.

By evening we reached Sawback Lake, which was our destination and where we were to camp for nearly two months. We had been following a stream all day, the horses picking their way through rocks, around boulders, and humping over the larger fallen logs. Suddenly the sound of rushing water became loud and immediate. We worked our way around a rock slope which barely left room for our packhorses to squeeze between it and the edge of the stream that cascaded in a waterfall. There just ahead, where the water poured through a logjam, was the outlet of the most dazzling little lake I had ever seen. The water was of the clearest green reflecting the sunlight as a wayward breeze chased tiny riffles across its surface and opalescent streaks of irregular pattern extended from the base of the mountains at the opposite end, where milky streams from their extensive snowfields met the lake. On both sides of the water lightly timbered slopes gave way to grassy meadows at timberline, two hundred or so feet above the lake, forming a basin that narrowed to an impassable barrier except for the narrow passage through which we had entered this magic little world. We had only to drag a log across this entrance, prop a few branches against it, and our horses were confined and safe with ample feed, good water, and shelter in the timber.

Belmore, who was in the lead, turned in his saddle, took off the "Mountie" hat that was his trademark, and his bald head gleamed as did the wide smile on his face. My expression of wonder and delight must have satisfied his greatest hopes, for he had chosen this special campsite for my introduction to the mountain wilderness of the northern Rockies that he knew so well. It was a place where there were fish and game species aplenty, rocks and snowfields to climb, and for him a nearly inexhaustible variety of scenes to paint—a secret place offered as a gift to share.

We set up the tepee at the upper end of the lake near the clearest of the little streams, at the edge of a lovely meadow. Soon the horses were loose and belled, and only Buck (who to his credit had caused no further trouble) was hobbled for ease in catching. I gathered wood and noted a fine, dry snag close to camp that would furnish wood later when it was felled. We finished the chores and started a small fire just as the sun dipped suddenly behind the western rim. At the same time a full moon that had floated unnoticed like a pale, transparent disk, began to take on substance. As it brightened and the last alpenglow faded to gray, the lake, the ridges and rock faces were bathed in shadowless light while the snowfields were silver.

We sat on a log we had drawn up to our fire, not speaking, each feeling and thinking his own thoughts, when quite suddenly the quiet surface of the lake seemed to be erupting as fish began to rise, some rolling to the surface leaving a swirl of brightness, and some leaping clear of the water in a shower and splash that caught the moonlight. Obviously there had been an insect hatch. Looking about us we saw that the air was filled with the fluttering white bodies of moths. It was on this clumsy bounty that the fish were feeding so ravenously. I do not think that even Belmore had seen such a sight.

We forgot our camp routines and rushed to assemble our fly rods. I had a number of white millers on small hooks, which we both tied on. We fished from the bank where the little stream flowed into the lake. Sometimes the fly was struck even before it hit the water. Each fish, when held to the firelight, was as emerald green on sides and back as the lake had appeared in daylight. They were nearly uniform, twelve inches or more, firm and full of fight, but obviously totally unacquainted with artificial lures. We caught about twenty in less than that number of minutes and continued fishing but throwing back our catch. Then just as suddenly as the fish had started feeding, the roiling of the water ceased, and although some moths still fluttered, no fish rose.

It had been like a frenzy for us as for the fish, and although I was always able to keep us in fish for the weeks of our stay, some days it took me hours to catch our daily quota, often resorting to grasshoppers and grubs as bait. I have fished lakes, streams, beaver ponds, and rivers in many different places and climates, but never has there been for me another such night when the churning emerald fish rose to meet our flies in one of nature's most felicitous conjunctions.

The next morning we felled the snag. Daily, after completing my chores, I was free. I had only to tell Belmore approximately what direction I would take, whether afoot or on horseback, and to adhere strictly to the rule that I must be back in camp in time for supper fire-making. After breakfast he would set up his easel and folding stool and prepare his palette. Squinting through the cigarette smoke that rose nearly continuously from his hand-rolled cigarettes, he would begin to paint. He did not seem to mind if I stood at some distance behind him and was quiet.

Belmore painted exactly what he saw, even though to some others the composition might be improved by liberties with the subject. But he could not move a tree or a rock, for the rocks he painted were of such substance and weight that any climber would trust his safety to them, and his trees stood so solid with rough pine or aspen smooth bark, that the experienced could almost feel. His towering cumulus clouds were water and lightning laden, and the filmy wisps of vapor that he sometimes caught as they clung to an edge of a pinnacle were ephemeral. When he painted an animal, it was not as a statue but arrested in a pose of alertness, motion, or peacefulness that held the essence of the species.

Only once did I see Belmore deviate from his precise

Belmore Browne painted rocks and trees with such weight and substance that any climber could trust his safety to them.

and literal approach to painting. I had heard him say before, "When I need money badly I can always paint a leaping trout and sell the picture to a banker." In any case, one day he had set up his easel by the outlet of the lake where the water cascaded into a deep, dark pool. He had painted the background of pines and sunlit openings, and before him he had placed one of the beautiful green trout from the lake, arched with mouth open, and erupting from a swirl of water to snatch a hovering fly. At that moment, incredibly, a Mayfly dipped and danced across the canvas and was captured by the wet paint just above the gaping jaws of the fish. Without change of expression, Belmore reached for a clean brush, dipped it in clear lacquer, and delicately entombed the fly. Never has a Mayfly been so beautifully reproduced. If some banker who owns a leaping trout by Belmore Browne should by chance read this story, let him examine the Mayfly in his picture and see if it is reproduced with the perfection of nature.

Belmore had experienced and loved the wilderness as few men have. He generously taught me a small part of what he knew, making me feel a scope of what was to be learned, seen, and felt but never fully comprehended.

Sometimes I rode out of the basin taking, at Belmore's insistence, one of the old horses that could be relied on to return to camp even if I became confused or ran into unforeseen difficulties. I had merely to move the log from the entrance to the basin and miles of valley, lower hills, and accessible alpine meadows were at hand. On these expeditions I saw my first wolf, grizzly bear, and woodland caribou.

I came on the wolf, a very dark specimen, standing watching my horse and me from the edge of a tongue of small lodgepole pine that projected into a marshy meadow. I looked down on him from the hillside less than a hundred yards away. He stood quite still, tail high, then threw back his head and gave a short, quavering howl. My horse had picked him out before I saw him and kept her ears pointed in his direction, snorting softly but not really alarmed. After a minute or two of mutual observation, he turned and vanished into the timber. We occasionally heard wolves howling, but I never saw one again.

My horse also discovered the grizzly bear as I was riding up a fairly open draw with a gentle breeze blowing in our faces that must have carried the frightening bear scent to her nostrils. She stopped, started to turn, and I could feel her back hump up and shiver under me. Being a good horse, she responded to heels and rein, and proceeded, with ears sharply pricked forward, taking tentative, skittish steps. Following the buckhorn rear sight of her ears, I now saw, about two hundred yards ahead and slightly above me the rounded, slowly swaying form of a huge grizzly bear. He had not seen us and was systematically turning over rocks and sniffing at groundhog and squirrel holes.

I was terrified and wondered whether to take the suggestion of my more experienced old horse, which was to scramble pell-mell over rocks, screes, through jumbled timber, back to the valley, or to sit quietly, watch, and hope the bear would pass above us, never knowing of our presence. I do not know whether it was paralysis, presence of mind, or prescience that triumphed, but we held our position, scarcely moving. The bear began to graze like a cow as he waddled through a patch of bright green grass. He swung his head shortsightedly and tested the wind, but still moved gently in our direction. He was angling from above so that if he held his somewhat wandering course, he would cross our back trail a hundred yards or so below and behind us.

I had no camera on this trip. I am not sure that I would have chanced the click of the shutter anyway, but I studied with more than academic interest the short snout, wide bulging forehead, humped grizzled shoulders, and powerful barrel-shaped body. His coat was still fairly smooth although it was well into June. As he walked with his great clawed front feet turning in with the awkward, swinging shuffle of plantigrade mammals, the hair seemed to ripple along his flanks and thighs. Somehow it appeared incongruous to watch this fearless and fearful predator working over the mountain meadow in search of beetles, grubs, and if lucky, a ground squirrel or mouse, for he was capable of pulling down horse and boy should the mood dictate. In any case, he passed us by at not more than fifty yards as we held our pose like artist's models. We could hear him rumbling and complaining when he passed behind us. An occasional rattling of stones indicated his progress when he was out of sight, then a moment of silence when he must have crossed our trail. We remained about two hours in the clearing through which he had passed, while I sat on a rock, drinking in the smell of warm earth and grass and regaining my composure. The horse grazed, often raising her head and watching the back trail. We returned cautiously at the end of that time to camp, seeing no further signs of bear. That night we stored a larger than usual pile of Indian kerosene with matches nearby in case we had a visitor. When one puts a match to a pile of dry pine and fir needles inside a tepee, it bursts into brilliant light, which is guaranteed to start the boldest bear for the other side of the mountain. We stayed awake for awhile, and Belmore told me stories of his many bear hunts in Alaska and of his killing the record brown bear that had carried off a full-grown caribou bull as though it had been a mouse in a cat's jaws.

On many days I went on foot across the meadows and on upward through and across the ice fields to the grassy slopes where a small band of sheep made their more or less permanent home. The band consisted of about twenty Rocky Mountain bighorn ewes with their lambs and a few yearling rams. At first I approached the sheep cautiously and patiently from below but found I had little luck in getting close to them. As I became more clever at the game, and the sheep became more used to my presence, I discovered the technique that most more experienced people already knew, locating a certain rock or projection on the hillside at a point slightly above and beyond the band and then, dropping completely out of sight and following some depression in the land, emerging at that point without my seeing the sheep or their seeing me. Day by day, we became accustomed to one another, and their tolerance of me increased to a point where I was able to come within thirty feet of the grazing or resting band.

Often I would sit with my back to a rock with sheep all around me and watch one particular ewe as she gazed fixedly with her yellowish eyes and chewed her cud with that rotary, rhythmic, mesmeric movement of the lower jaw characteristic of all ungulates. I would watch the cud roll up her esophagus like a golf ball, be chewed for a minute or two, and then roll back down into the reticulum, although at that time I did not know much about the anatomy of bovids. From time to time a ewe would rise, stretch, urinate, and start walking slowly away from her bed toward the bunch of lambs that were usually playing at some distance. She would utter a soft bleat, and almost immediately her lamb would break away from the group and run toward her, ducking under her flank, grasping one of her two teats, and with vigorous butts that almost lifted her hindquarters from the ground, start feeding while its little tail waggled frantically in the way of all nursing sheep.

Two mountain goats also frequented the higher reaches of this Sawback Ridge, but I was not successful in approaching them closely, even though I attempted the same technique as with the sheep. When I arrived at the point where I expected to find them, they were already above and ahead of me so they were usually impossible to approach from above. I was able, however, to come close enough to watch their extraordinary climbing techniques and abilities and compare them to those of the sheep.

The sheep, when traveling at speed, jumped from point to point with graceful arching leaps, usually landing with their four feet together if the terrain was rough. On the few occasions when I saw them really alarmed and starting downward over a steep slope where a fall would be fatal, yet where no ledges or projections were wide enough to permit a landing, they would fearlessly drop over the edge, striking even the smallest projection with bent knees, thus breaking their fall, then on to the next tiny ledge until they reached the bottom safely. The lambs negotiated any terrain that the mothers chose. Steep, nearly vertical upward slopes, however, were not attempted by the sheep; they chose circuitous routes around the excessively steep places.

The goats, on the other hand, showed far less panache in their mountaineering techniques, traveling slowly and deliberately as a rule across the most difficult terrain. Like human climbers, with feet and hands, they usually kept a secure foothold for three of their four hooves at any one time. When climbing a nearly vertical surface, they would stand on some narrow shelf, reach upward with their front hooves, if necessary nearly as high as their muzzles, bending the fetlock at forty-five degrees, and chin themselves until their hindlegs could reach the level of their front foothold. Their hooves were peculiarly adapted to this technique, having sharp edges and soft, rough, rubberlike pads between. Their knees are placed low on the front legs, making this chinning activity possible.

While watching the sheep, particularly in the evening, I often saw a fine mule deer buck, his antlers still in the velvet and not fully grown. He would emerge from a patch of timber and move into the forest edge, quite near the lake, for his evening foraging. There were also two mule deer does, one probably with her first fawn, and the other, perhaps her mother, with twin fawns, and they, too, came to the meadow in the evening. There were marmots that whistled their alarm at me from the lookout rocks above their dens, where they apparently did not tolerate crowding from other members of their species. There were also the ground squirrels, and in the woods the grouse, locally called fool hens, because they did not fear man. They offered the most tempting of targets for stick or stone for anyone tiring of fish, bacon, beans, and dry meat.

On one occasion I took time to cross the Sawback Ridge by a quite easy route and dropped over the other side into a more heavily timbered valley. As I entered the edge of this lodgepole pine forest, I could see in an opening a magnificent mountain or woodland caribou. I had never seen a caribou before, and this fine bull bore little resemblance to the reindeer that is a pale replica of this wild cousin. He had shed out his winter wool, and his nearly black coat gleamed over the rolls of fat he had accumulated during the abundance of spring and summer. His neck was draped in a magnificent white mane and collar that extended from jaw to breast, and his enormous antlers with the right brow tine shoveled, reached nearly to the tip of his furred muzzle. They were still in the velvet, adding to their massive appearance. No other member of the deer family has proportionately such great antlers as the caribou compared to its size. I would judge that this bull must have weighed close to three hundred and fifty pounds. He apparently heard my approach, turned to face me, but did not seem alarmed, and we watched each other for a moment or two. Then, with a long, swinging trot and the clicking ankles characteristic of the species, he moved out across the clearing and disappeared into the forest beyond.

I realized it was late and I had far to go, so filled with excitement I started back up over the ridge, down across the rocks and snowfield and back to camp, a little late for fire-making and my evening chores. But when I told Belmore of the caribou, I was forgiven. He said that he had seen many caribou in the Alaska Range, in the foothills and lower elevations of Mount McKinley, but these caribou traveled in small bands, were often found at elevations and in terrain as difficult and as high as the sheep, and could negotiate this rugged terrain nearly as well. He also commented that he had frequently seen cow caribou bearing antlers as late as April. (I do not think it was generally known then that pregnant cow caribou often do retain antlers long after barren cows and bulls have shed theirs, presumably providing competitive advantage.)

In the evenings after supper when the dishes were washed and the oatmeal bubbled on the back of the fire, we would sit and watch the light fade until it was almost dark. We made sure the ax was properly buried in the chopping block, the pots and pans turned over, and all food out of reach of large and small prowlers of the night. Then we moved into the tepee, sat on our rolled-up sleeping bags, and lit the tiny fire in the center of the tepee. We had no lantern, only some candle stubs and a flashlight for emergencies. We brought no books, nor needed any, for we talked until it was time for sleep, or rather Belmore talked and I urged him on with questions and occasional comments. As I listened I fed small twigs into the little fire and the flames leaped up, lighting the tepee, which reflected the flickering flames from its white walls.

Most of all, I wanted to hear about his three expeditions in the Alaska Range and his final conquest of Mount McKinley, and he seemed willing to relive this period for me. I will not pretend that I remember in detail or in proper sequence his narration, but I have reread his magnificent book, *The Conquest of Mount McKinley* (G. P. Putnam's Sons, Knickerbocker Press, New York, 1913). It is a large, beautifully composed book, illustrated with more than a hundred drawings, paintings, and photographs of his three incredible adventures. My rereading of the book has served to recall those evenings in the tepee listening to his precise but pleasant voice tell of deeds of almost unbelievable endurance, resourcefulness, and courage. Belmore's expeditions were not undertaken in a search for glory, but with a high-hearted hunger for adventure and a call to penetrate unknown valleys, cross unmapped ranges, and finally, if fate, human fortitude, and careful logistics all conspired (which they did), to reach the unclimbed summit of Mount McKinley, the great mountain of the North American continent. I shall never forget the impact of these stories, and all my life I think my concept of things being "really rough" has been profoundly modified as a consequence.

After an idyllic two months in the mountains it was time to head for home. The nights were getting cooler; often there was frost on the meadows in the morning and school would soon start.

Buck had been turned out with the rest of the horses, but I had hobbled him and left a rope dragging from his halter because he was often hard to catch. The day before we were to leave, I checked his fetlocks as he had been wearing the hobbles for a long time. They looked a little puffy, and I decided to remove the hobbles and picket him until we were ready to go. I had no trouble catching him by the slimy, tattered rope he had been trailing, and removed his stiff, awkward hobbles while holding the rope. Something must have spooked him, for no sooner were his hobbles off than he pulled back and was gone, trailing the rope and heading for the timber at a thundering gallop. I followed him quietly, talking and soothing. I tried all the wiles and tricks I knew. Belmore did and said nothing. I tried to catch him by approaching on horseback, but to no avail. He kept his distance, head high, snorting when approached. We thought that since he was well integrated into the horse herd he would probably follow us out the next day, even if at a distance, but he did not, and we could not delay our departure. I, frustrated and humiliated, mounted the pale sleepy little roan with which I had started, while Buck, a glowing gold patch against the hillside, watched us light out for home. Perhaps he joined some other pack string as winter closed in, and someone else became the owner of a big, rawboned, Roman-nosed, splay-footed, ornery buckskin gelding with sensitive ears. Or perhaps out there in the more protected valleys of the Alberta Rockies a lean, hard-bitten buckskin horse, his halter long rotted away, survived the mountain winters, and fattened again on the rich, spring grasses for as many years as are allotted to a strong, Indian pony. I had set out proudly, like Don Quixote on my tall, big-boned "Rosanante," and was returning wiser, more like Sancho Panza on "Dapple," my humble little roan.

Exploring the Northern Sierra

Summers were always pleasant in Santa Barbara, and since jobs were nearly impossible to obtain, we passed our vacations in various ways. I enjoyed the beaches and also began, somewhat later than my friends, to engage in a relaxed and delightful social life with a group of friends of both sexes, many of whom were struggling young artists who had come to Santa Barbara attracted by the excellent art school. Our lives revolved around two stabilizing centers—my mother and Tom's tea table, and the studio of an artist, Dudley Carpenter. His quiet warmth and wisdom invited confidence and provided perspective during those growing-up years whose turbulence, despair, and occasional exaltation could not be and should not have been entirely tempered.

In the middle of Dudley's big, white-walled Spanish studio lay a large, faded cinnamon bear rug with head attached, teeth missing, and skull soft from much hard use. Many were the evenings we sat around the small fireplace on this rug with the chunks of live oak glowing, music playing softly while we talked or sketched, or planned, or dreamed, or argued, calling on Dudley for quiet comments and soft judgments. Around the tea table, on the other hand, the conversation tended to be less personal, more intellectual, and far more cosmopolitan and wide-ranging.

Yet I was always restless by the end of summer and sated

with the social and civilized ambience of Santa Barbara, feeling the need to get out into the brush again. Fortunately, I had a friend, Dick Hanna, whose family owned a beautiful cattle and horse ranch in a valley bordering on the Mount Lassen National Park near Mineral. It was green at all times of the year, even when the surrounding country was deep in snow, for the mile-long valley floor was dotted with steaming hot springs. The springs warmed the ground that lay between abrupt mountainsides and rimrocks timbered in some areas and clothed in nearly impenetrable thickets of chaparral in others. Down the center of this valley a cold swift trout stream—Mill Creek—flowed between the meadows and on toward the Sacramento River through a steep cut in the mountains. The ranch house was a simple but gracious one-story building with a long, shady porch that looked over the well-kept hay and stock barns in the distance. A fine herd of purebred Hereford cattle grazed on the fertile upper meadows, while across the road in the wetter meadows, golden horses fed contentedly. These remarkable horses were being bred from a stallion and a few mares of the same bloodline, which had been carefully traced and acquired by the Hannas. Their colts seemed to be consistently gold colored, but unlike palominos, their manes and tails were the same shade as their bodies. I wonder what has happened to this strain, as I have never seen such horses elsewhere.

It should not be hard to imagine that a visit of two weeks or so to this valley in early September was superb. We rode through the surrounding mountains accompanied by two Chesapeake Bay dogs, a mother and son that knew more about deer than we would ever learn, and prepared for hunting season by spotting where the big bucks were bedding high above the rimrocks. When opening day arrived we started early in the morning and rode to the cliff tops. After tying our horses, we sat down quietly at our favorite lookouts while the dogs worked the brush patches below. They worked silently through the stiff, prickly undergrowth, with typical Chesapeake toughness and apparent insensitivity to minor pain. We watched the edges of the brush, and if luck was with us, first a doe would move out ahead of the dogs, traveling close to the ground from cover to cover until out of sight. Then perhaps some minutes later a buck might come busting out with great mule deer bounds. There was just a moment for a snap shot that mean success, or more often in my case, a miss, but usually by the second or third day of the season we each had our deer. In the evenings we fifteen- and sixteen-year-olds took the pickup to the little town of Chester and the illicit bar where a semitame bear was chained just out of reach of the customers. There we bragged of our hunting prowess while drinking sour prohibition "Angelica" wine.

Many San Francisco hunters came up during the weekend of the deer hunting season and often failed to bag a deer and were desperate about returning empty-handed. An old goatherd name Matlak took advantage of this. Each year he supplemented his income derived from angora goat wool by bringing some of his herd up to a rimrock that rose nearly vertically from the road edge where these doughty hunters passed. He drove a few of his oldest cull goats out onto the rock face where their white coats made them conspicuous, then sat like a spider, patiently awaiting his prey. Sure enough, several times a day during the season, a disappointed city hunter driving the road would see the goats, bring his car to a screeching halt, look about guiltily, notice that he could stand unseen from the road behind a small clump of timber near the foot of the cliff, and fall headlong into temptation. With one easy shot he got himself a goat, thinking that when skinned and dressed out it would suddenly become a deer. At this point the herder stepped out with pretended fury, claimed a huge compensation (like fifty dollars), which the embarrassed and demoralized hunter gladly paid. Then with a wide, toothless smile, Matlak skinned and dressed the goat and shook the hunter's hand, which required a closeness of contact that assaulted the latter's nostrils with the blended stench of billy goat and unwashed man. He made a lot of money that way.

After fall hunting on the ranch, we usually stayed on to help with chores: restocking the woodpiles at the three campgrounds that were maintained for guests, repairing rail fences that protected the cattle from the hot springs, working stock, and with buggy, buckboard, or wagon, doing any hauling that was needed on the narrow dirt roads where trucks could not go. It was small repayment for the magnificent hospitality, and it was my first real opportunity to be around beef cattle and ranch life in general. One day, once the chores were under control, I took my fly rod and started down Mill Creek. This lovely, cool stream originated high on the southern slopes of Mount Shasta as a rivulet fed by the mountain snows, and by the time it reached the ranch, swollen with the numerous small tributaries, it had become a swift but friendly, fordable, and eminently fishable stream. I had asked some of the ranch hands what the creek was like for camping and fishing after it entered the wilder hill country some five or six miles below us, and I also consulted the neighbors at Childs' Meadows. There was a curious reluctance to discuss the matter. The general reaction was that the country was impassable, that there was no good fishing, and, in fact, that the Mill Creek canyon and gorges were dangerous to travel. As far as I could find out, the creek had never been surveyed throughout its entire forty-mile course.

From the local forestry maps it was evident that the area between Deer Creek, running parallel to and some ten miles south of Mill Creek, and the old Shasta Trail (used in the 1950s and 1960s but now abandoned), was without topographical details from Mill Creek's entry into the mountains until its emergence onto the plain. In 1929 there were few such blank areas left on the maps of the United States, making this stretch of creek, perhaps not more than fifteen miles in length, a mystery that I felt needed investigating.

There was something about this blank corridor on the map that haunted me—so well-known at both its upper and lower reaches, but sinister and shunned in the middle section where it plunged into the broken mountains of the range. I really wanted to get through that canyon and finally persuaded my closest school friend, Norman Trevelyan, to join me. I was honest about the country and the problems. He was game, and we planned carefully, agreeing that if we could not get through the canyon to the valley in six days we would have to turn back and that we needed ten days' worth of supplies to give us a margin of safety. I used my Aleut pack strap, copied from Belmore's design, while Norman had a good, light mountain pack. We started out each carrying forty pounds. We took some rope with us, but otherwise included few extras and no luxuries. We knew that we could not count on fish much after the first day.

We started down Mill Creek early one morning after a huge breakfast, completing the first five miles of the journey and catching a good mess of nine- to twelve-inch trout before making camp. This would be the last night of the trip when we were able to sleep on soft soil and cook with abundant dry wood.

Well-fed, we started off just after sunup. The going was fairly easy for about a mile, and then the mountains began to close in. The shore between the increasingly swift stream and the cliff edge narrowed until more and more often we had to jump from rock to rock to circumvent a spur that jutted out into the swirling current. The pebbles and small rocks by the water's edge were drenched with spray, and the larger boulders were slippery and treacherous. Soon it became impossible to follow the streambed for more than short distances. We had to turn to the rock walls to find uncertain pathways where roots of moist underbrush and fissures or protrusions furnished foot and handholds. If lucky, we did not have to go all the way to the top of the cliffs, but found shelves running more or less horizontally along their faces. These ledges were often overhung with rock, forming caves and, if wide enough, supported a few bay trees and conifers in their shallow soil. Then we would have to go down again to the creek bed, having progressed perhaps less than a quarter of a mile downstream for the two miles of near vertical climbing and descent. It was like this for two more days, except that the bluffs rose higher and higher, until we felt trapped and squeezed into a slit with only a tiny bit of sky directly above our heads. It was twilight even at midday.

To say that we were discouraged puts it mildly. By the evening of the fifth day, our packs seemed intolerably heavy as we climbed nearly vertical rock walls only to find dense chaparral at the top. Progress became a torment of scratches, slaps, and clinging angled branches through which one could have only traveled upright by chopping a path with a machete that we did not carry. We crawled under the worst of it and kept close to the edge, constantly peering down to find a route back to the canyon floor where a wide enough shingle for a campsite might be found. Finally, we came upon a pebbly little beach, about fifteen feet long and three wide.

We estimated that we had come fifteen or sixteen miles from the ranch, the last ten miles of which were in the unmapped gorge—five or so more to go before we broke out into the open. We were exhausted, and more than that, we were depressed in a way I, at least, had never been. The dangerous, thundering creek foamed between smooth boulders, curled and sucked through narrow passages, roared over falls, and made the tiny, moist patch of gravel upon which we were dependent shake constantly with the pulse of the torrent's rhythm. Worst of all, the thunder of the water reverberating between those immense rock walls assaulted the ears and made our nerve endings tingle with nearly unbearable irritation. This sinister wound cut deep and, revealing sights and sounds from which the ear and eye are usually shielded, seemed to have become a veritable Coleridge fantasy, dreamed up in a laudanum nightmare and called "Alph." The tumult was so overpowering that we could not make ourselves heard even in shouts. We only looked at each other with thinly concealed animosity—we who had always been close friends. Still, we had to make the best of it, and clearing away the more uncomfortable rocks, made a small fireplace, and scrounged around on the cliff-side for what dry bits of wood we could find.

It was at this point that I became conscious of a movement just downstream on our little slit of territory. A porcupine had been hiding a few yards away behind a rock. It approached us, rattling its quills, switching its tail, and obviously grumbling or uttering the babylike cry of lovesick or disturbed porkies, although we could not hear it. Perhaps our fire had finally driven it to try to escape. To reach the little trail down which we had come, it had to cross within a few feet of us. It was obviously not going to give way, and yet there was not room to pass. I had a twenty-two revolver in my pack and shot it at a distance of about three feet before the inevitable confrontation. Then, knowing that in the north one never destroys a porcupine needlessly, as it is the only animal that an unarmed man can obtain for food, I turned it over to dress it out. The prospect of fresh meat seemed appealing. As I made the first long cut from groin to diaphragm, I realized that it was a lactating female, and the sticky milk spread over the entire carcass. Ordinarily this would have been no great shock to me, but when I saw Norman's face, I realized that porcupine was not on the menu tonight. We ate bacon, biscuits cooked in the light frying pan, dried fruit, and coffee; then, without communication, tried to arrange two sleeping bags on the impossibly stony, narrow, and moist shingle and attempted to blot out the roaring and quaking of the water and the earth by covering our heads in the sleeping bags.

Neither of us slept much. After a glum breakfast, Norman motioned upstream with the thumb in the typical hitchhiker's gesture, and I nodded. Silently we retraced our journey, but with slightly less difficulty, as we now knew where to circumvent the obstacles. Often traveling looked as though it would be easier on the opposite side of the creek, but it would have been madness to risk a fall from a slippery boulder into that churning current unbroken by logs or other natural crossings. It took us three days to get back—only eight days for the whole trip—but we both felt as though we had glimpsed something outside of ordinary experience. It was wildly beautiful, but a grotesque rendering, with more than three dimensions. We both felt that if we looked quickly over our shoulders, we might almost glimpse shadows, almost hear silent treads. Neither of us is much given to fantasy, but we sensed a presence that we never mentioned.

In 1968 my brother-in-law sent me a book that he was sure I would enjoy, although he knew nothing about my Mill Creek experiences of some forty years before. The book tells the extraordinarily moving story of Ishi (meaning "man" in his native language, for he would never use his Indian name), the last stone age Indian to emerge into the twentieth century. Ishi was the sole survivor the Yahi Indians, a once proud and isolated tribe in the remote gorges of Deer and Mill creeks in the Northern California mountains, which was nearly exterminated in the 1800s and went into concealment in 1872. It was assumed that the nation was extinct, for there were no raids on lonely shacks, no smoke seen curling thinly through the hot skies of summer or from the deep canyon floors, no broken arrow shafts, and no telltale tracks. But one small family group, including Ishi, survived for a time, in the fearsome gorges of the Mill and Deer creeks. They led a furtive existence, hunting and fishing with harpoons and milkweed nets, and hunting deer and small game with their finely made bows and feathered hazelwood arrows. Finally, there was only Ishi, who, half-crazed with loneliness and hunger, desperately revealed himself to his tribal enemy, the white man.

Ishi is a remarkable book, beautifully written and meticulously researched by Theodora Kroeber. It was published in 1962 by the University of California Press, fifty-three years after Ishi's death, thus few people during earlier years knew anything about his story. It would have been of deep interest, especially to the few of us who had entered the secret Yahi world of the Mill Creek gorges, quite unaware that the overhanging ledges, the sheltering caves, and the hidden tunnels through the chaparral, had made possible the nearly total concealment for almost twenty-five years of a small group of aboriginal people. We did not look for evidence of Indian occupation and found no broken arrows, no clay-lined baskets, no bones, although it had only been twenty years before that Ishi's people had traveled with far greater ease and skill the course we followed through this forbidding chasm.

Dirt Roads and Rodeos

THE SUMMER FOLLOWING our Mill Creek expedition should have been my last before college, but because I was only sixteen Mr. Cate suggested that I stay on another year to help with the camping program and to finish out my college board credits in a more leisurely fashion. I endorsed the plan and decided to spend this extra summer of 1931 driving to Chicago to visit my grandparents and other members of the family. It was not to be a fast or direct trip, and the character of the expedition changed somewhat when my friend Bud Walker wanted to join me.

First of all, we decided to drive the 1927 robin's egg blue Chevrolet roadster with the rumble seat that I had inherited from my brother. The car was in good condition allowing us to make room for camping gear, guns, and food by removing the rumble seat cushions. We then attached two ten-gallon wooden water kegs and extra gas cans to the running boards. After we each contributed the munificent sum of two hundred dollars to the common fund, we established the goals and purposes of the trip: to follow the rodeo circuit when practical, never sleep under a roof except when visiting friends, and provide all our own meat by hunting along the way.

We were off, heading north, the little roadster packed to capacity. Our first major provisioning was at the Hanna ranch near Mount Lassen. As the owners were away, we camped at the upper end of the valley and spent three days jerking enough venison to carry us through lean times. Then, drifting north and slightly east, we took in the small and large rodeos in Pendleton, Oregon, and Ellensburg and Wenatchee, Washington, and finally Wyoming, where the season ended. Sometimes we camped close to the fairgrounds where the rodeos were held, but more often we found pleasant campsites away from the towns.

The smaller rodeos with competitors mostly from the local ranches were always more fun for us. They were informal, the bronc riders used their own working saddles, and often broke a bottle of orange pop (when nobody was looking) over the unpadded seat to provide a good sticky surface during the ride. Lariats were forty to fifty feet long, made of manila or braided rawhide instead of the stiff, synthetic fiber of today. Cowboys threw them with "wildcat loops" rather than the small rigid loops clapped over the heads of calves in the rodeos of today. Despite prohibition, booze flowed freely, which probably resulted in more frequent falls, although broken bones were perhaps minimized by the more relaxed state of some of the competitors. Often these rodeos held calf-riding events for the kids. A ten-second ride was worth ten dollars. I enjoyed modest success in calf riding on several occasions, learning to compensate for the amazingly loose skin of a six- to eight-month-old calf, which readily slides from neck to rump, from spine to belly, and back again. I sometimes suffered bruising from a piston-like kick of a calf as I dismounted, either at the ten-second mark or before. Since then, I have had many opportunities to learn respect for the sage dictum: "Keep your belt buckle up against its ass when working stock in a chute."

We moseyed our way across Washington, mostly on the more primitive of dirt roads, and into the Idaho panhandle. Camping was delightful. We could always find a stream to fish and prairie chickens for meat. Jackrabbits, as far as I am concerned, are inedible, yet they were all around us, especially in the evenings as we drove those nearly deserted roads.

They offered an almost irresistible temptation for target practice. Except for jackrabbits, we shot only for the day's supper as it was summer and no ice was available anywhere.

It had been a spontaneous trip until we entered Canada on our way to Banff, where I wanted to show my friend a little of that Canadian Rockies country about which I still dreamed. For the most part, the narrow "main road" leading from Kingsgate to Cranbrook was surfaced with round river gravel so that cars slipped and slid on the sharp turns. About twenty miles from Cranbrook the road ran along an outside curve on the hillside, perhaps fifty feet above a creek. A power line sat on a narrow shelf about halfway down from roadbed to stream. I misjudged the turn, started to skid sideways toward the edge, and attempting to correct, swung the wheels toward the drop, and we sailed off into space. Had there not been a pole directly in our flight path, we would have ended our short careers, but instead, we hit the pole squarely at about the halfway mark, slid down it, and came to rest on the narrow ledge with all the tires flat, the radiator collapsed, the windshield broken, but the motor still idling away.

We were shaken but quite unhurt. We sat trembling and sweating, then finally turned off the motor, which was about to blow up from overheating, and surveyed the damage. We climbed to the level of the road, and looking back, noticed that the power pole and line were not seriously damaged, but only tilted outward—it seemed miraculous. There was nothing to do but sit by the roadside to wait for help. It is hard to believe that only fifty years ago the major roads, at least in the West, were little used except by light local traffic. Even the "Lincoln Highway," running from Salt Lake City to Cheyenne and Omaha and on to Chicago, now designated Interstate 80, was unpaved over many miles of construction detours, some of which were nearly impassable gumbo mud after sudden rains. Therefore we expected few cars on the Cranbrook road, and it was about an hour before the first vehicle could be heard, grinding its way up the hill.

It was an old truck, and the farmer who drove it was hesitant to stop on the hill, but finally consented, waiting for us to place rocks behind all four wheels before turning off his engine. He was unfriendly and disbelieving when we told him our car had gone off the road, but at last he got out and looked over the edge. When he saw the car with California license plates, which he had never seen before, he shook his head in disbelief. Reluctantly, he agreed to drive us on to Cranbrook. We struggled back down the slope to get what gear we needed until the car could be retrieved. It never occurred to us that the contents of the car would not be safe, and although the car remained on the ledge for two days, nothing was purloined. Finally the Cranbrook garage owner obtained a sufficiently sturdy wrecker to haul the little roadster up to the road and tow it in on two new tires.

There were no parts for a '27 Chevrolet in Cranbrook. After wiring to Calgary, we were told that it would take ten days to get a new radiator and whatever else we needed. So we were stuck in what was then a small provincial Canadian town, short of funds for such a disaster, and obviously unable to carry out our original commitment to the outdoor life. Accordingly, we took a little room above the tavern (no prohibition then in Alberta) and settled down to wait. Each afternoon we visited all three churches in turn, and every evening we went to the movie. I shall never forget that movie. It starred a very boyish Robert Young and a high-schoolish but captivating Ginger Rogers in a story about the navy. They were both in sailor suits, and there was much dancing, singing, and romancing of the most wholesome type. That picture saved us from utter boredom.

At last the car was fixed—that is they rendered it operational. We dispensed with the top and windshield, a mixed blessing, for although large grasshoppers and hard-shelled beetles occasionally caused stinging blows to our unprotected faces, the shooting was much simpler with no glass in the way. As to the top, when it rained it was no different from being outside, and the sun was hot, even with the old canvas top. But more importantly, such luxuries were unaffordable.

We were off again, briefly passing through Banff, where much to my chagrin I found that Belmore was in New York. But we camped for two days at the Brewster Stable before turning south, now accompanied by a little six-week-old Husky puppy we had bought from the government kennel. We were headed for Coeur d'Alene, Idaho, where a school friend, Tom Greenough, had offered us lodging and a good time as some friends of his sisters were visiting from a school in Massachusetts.

We decided, because of possible problems bringing the pup across the border, to try a crossing that showed only on a local map. It was little traveled, and luckily we met no one on the thirty to forty miles before it rejoined a main road, for passing would have been impossible. I shall not forget one ravine crossing bridged by two logs, perhaps twenty feet long and wide enough to accommodate the wheels of a car. The tops of the logs were roughly adzed to provide a slightly flattened surface, and we were relieved when we finally inched to the other side without accident. We found ourselves back in the United States not knowing when we had crossed the border. The poor puppy suffered greatly from the heat, so we wrapped him in damp sacking and fed him finely cut up pieces of whatever meat we had obtained for ourselves as we had no milk.

Coeur d'Alene is a beautiful town on one of the loveliest lakes in Idaho, and our friend's house faced the water. We were royally welcomed, cleaned up, laundered, housed and, as a bonus, entertained by three lovely and refined ladies from Miss Hall's School in Pittsfield, Massachusetts.

After about a week of this debilitating life, we took off for Missoula, Montana, where Tom's grandmother lived. There Tom introduced us to her marvelous old Victorian house of many rooms and turrets. The bathrooms had griffin-footed tubs of majestic proportions and commodes with varnished wooden water tanks high overhead and tasseled silk cords to pull. Such pulls created a most satisfying roaring, flushing action that announced to all one's location and activities. The furniture, pictures, and bric-a-brac were magnificent, fringed and gilded. The lawn was not neglected either; iron deer and a few rabbits were tastefully scattered over the velvet grass.

Best of all were the stables, which still housed a fine collection of carriages, sulkies, and a surrey, with horses to match. We were permitted to drive in style through the many acres of the property, now a Missoula city park. The old house has been moved recently and is now the property of the city, but at that time Tom's tiny, round grandmother was its mistress. She was plainly dressed with her white hair drawn back into an unpretentious bun but radiated a warmth, strength, and charm that I will never forget. For three days we were all her favorite grandchildren, and we loved her for it. I am sure that old-timers in Missoula who knew old Mrs. Greenough would respectfully acknowledge the role she must have played in the history of that city.

We covered many miles and paused at many campsites during the rest of our journey. We lived in much the same manner as I have described, traveling through country new to us. We went south to Albuquerque, still a smallish town, to fascinating Santa Fe, and up into the Sangre de Cristo Mountains, which were then as wild as any country I had experienced. The towns and cities of that region were like those of an exotic foreign land: vaqueros on their horses; Indians in blankets and/or "whiteman's" clothes; low adobe houses, dirt streets and the lovely sound of Spanish and its songs and rhythms. A precious heritage met us, and we lingered in its ambience before turning to the north.

We drove through Colorado into Wyoming, where we took in the Cheyenne Rodeo and best of all, the one in Sheridan. There we camped "by invitation" on the edge of a big Indian encampment and shared, a little reluctantly, the chunks of meat that the squaws had hacked from the carcasses of two bison the Indians had been given permission to pursue and kill in the old manner. The weather was stifling:

the buzz of flies drawn to the offal, the sweet smell of decomposing meat, Indian tepee camp, Hellyer camp, rodeo, stock, and manure, all lent their particular character to the festive atmosphere of the rodeo. In the chutes and the arena the classic western contests of man with horse, man against horse or bull or steer or calf, were waged in the dust. There were brave clowns, pretty girls on horseback, buckskinned Indians with feathered headdresses on paint ponies, and in the background, the drawling voice of the announcer, and the nasal, mostly sad cowboy ballads, quite unlike the country and western sounds of today.

It was but a short distance from Cheyenne into South Dakota where we found pheasants plentiful and delicious. Good camping sites were more difficult to find, and the heat persisted long into the night. We both became extremely sick one evening and holed up in a sweltering little gulch by a semistagnant pool with scant shade, where we vomited, shivered, had diarrhea, not caring about anything for two days. When it was over on the morning of the third day, we decided that we had had enough and climbed weakly into the car to start east toward Chicago. As the day progressed and our faces, unprotected by a windshield, turned stiff from sun and crusted from the dust, we decided to push on to my grandmother's home in Riverside. There we knew we would have wholesome food, a bath, a lot of good-natured teasing from Bob Randolph, and no care for the morrow.

We arrived in their driveway at about 6:00 P.M. and were met by Uncle Bob (as we affectionately called my step-grandfather). He pretended lack of surprise and said, "Here, let me help you damn kids unpack, and leave all these clothes by the basement stairs outside." He looked at the dented, windshieldless car with the missing right front fender and jerked open the rumble seat, only to recoil. We had forgotten to throw out two prairie chicken carcasses we had planned to eat before we became ill, and they had ripened to a slimy, half-wrapped mound on the floorboards.

I cleaned up the car and did the same with myself. Both

Bud Walker and I were quite content to surrender ourselves to the delights of civilization and grandmotherly ministrations. That evening my brother George and his new wife, Babbie, came over and I gave them as a belated wedding present the Husky puppy (now long-legged, friendly enough, and very handsome with his black saddle and spectacled face). After a few days of recuperation, Bud offered me forty dollars for the car as he planned to drive on to his home in New York. There he was able to resell it to a poultry farmer for twenty dollars. It had certainly given us its all that summer.

From West to East

AS SOON AS I RETURNED to school, I set about expanding and improving the camping program. Mr. Cate was willing to go along with my plans and even agreed to feed the two small mules we acquired. Although I had had no experience with mules at that time, I knew they were easy keepers. Somehow in that Southern California setting and Spanish tradition there was a picturesque appeal in their use as pack animals.

Enlisting the help of a few other camping enthusiasts, we acquired pack saddles and other gear suitable for small mules. Soon nearly every weekend four of us would be off, sometimes as early as Friday afternoon, with ample food and a dose of liquid contraband. We roved widely, both north and south along and behind the coast ranges. It was mostly rough travel, and each night we had to drop down into valleys to find water and horse feed. We carried no tents during the camping season as rain was infrequent, and a tepee would have been useless in a land where poles could rarely be obtained. In the evenings we had the luxury of a Coleman lamp, hissing from a cord slung from an overhead limb, and I, as cook, often concocted baked delicacies in a reflector oven and abundant meals from the extraordinarily good food furnished by the school from the list we submitted each Thursday. After supper we cleared off the tops of the pack boxes, and with backs propped against our saddles, we played bridge and poker and increasingly worried about college boards.

During this period of my life, however, when so much of my time was involved in outdoor activities, I had not sacrificed my interest in reading, particularly history and biography. I started a William James club that met weekly to discuss pragmatism (a word not in common usage then), as well as reading from the classical philosophers. Most of my friends were heading for Harvard, but Bob Savage, a member of our camping group, and I were applying to Yale. I did so because Tom Ripley (Yale '88) encouraged the idea. I also felt that an experience of the eastern United States would be valuable when the time came for job hunting in what was then the Great Depression.

Frequent camping trips helped to pass the senior year at the Santa Barbara School. The buckskin shirt went on every one of them.

We were expected to work hard at school, and I enjoyed learning, although I was never one of the quick ones to understand a new concept. I was and am of the Gestalt mentality, so that after dragging along half or even three-quarters of the way through a new subject, with undistinguished grades, suddenly—ah-hah!—everything falls into place and the subject is mine.

Thus this last year at the Santa Barbara School passed quickly, but like many high school seniors I recognized that a golden period was passing all too fast. I tried to cram all the outdoor activities I could into that year, for by Christmas I had fulfilled all my college requirements and was already accepted at Yale to enter in the fall of 1932. I realized that the opportunity for the free life on horseback in those lovely California mountains was coming to an end for me and probably for the school as well, for the land behind the mesa was up for sale while the surrounding dirt roads were rapidly being paved. As a speaker on that last commencement day, I made a plea that this land be acquired by the school, quoting Kipling's "The Explorer":

> Something hidden. Go and find it. Go and look behind the Ranges—
> "Something lost behind the Ranges. Lost and waiting for you. Go!"

A wealthy patron of the school offered to purchase the ridge behind the mesa thus maintaining access to the trails that led over the mountains and into the Santa Inez valley. But it was evident that the times they were a'changing. Many of the new boys had little interest in horses. Tennis, mechanics, basketball, and a modern gymnasium were on the list of needs. The money earmarked for the purchase of great brush hillsides, rock outcrops, and hidden pockets of green where small pools withstood the parched summer was turned to other uses. Teachers with interests other than the outdoor life were beginning to fill the ranks of a new and equally talented faculty. The school grew, changed, and now ranks in the forefront of private coeducational schools.

The backcountry is now privately owned and much of it developed. The Santa Inez holds a beautiful lake dammed up at a spot not far from where the little old stone school cabin stood, and fine ranches flourish where the little river once meandered through the scrubby brushland. No longer can a boy on horseback look up as the shadow of a California condor sweeps across the valley, a once common passage over terrain they used to share.

In a way, I must have been ready for this new phase of my life, for that summer I sold the fine saddle that had been made especially for me by Visalia's in San Francisco, together with most of my other horse and camping gear. I also sold my horse and mule and entered with enthusiasm into the social life of beach and parties, indulging freely my susceptibility to feminine charms. I began to anticipate the urban and academic life of New Haven and deliberately turned my back on the hills.

Two of my classmates, Bob Savage and Norman Trevelyan, and I decided to drive east in a car that Bob and I had acquired jointly. To call this chariot a car does it little justice, for it was no ordinary automobile. It was a 1928 second-hand Jordan Speedboy—a long touring car with a cowl and windshield that raised and lowered over the back seat to protect the passengers, red Morocco leather upholstery, an engine with an aluminum head, and an immense Stromberg-Carlson carburetor with, it was reported, fifteen jets. Over the next several years, however, it operated smoothly with far fewer jets. In several small-town garages where we were stalled at one time or another, the local mechanics took it apart. After putting it back together, they were left with several unused parts, yet the car would run as well as before their deletion. The Jordan had only one defect: it hated altitudes over three thousand feet when it would vapor lock. Neither we nor the various mechanics in the high country were able to make the diagnoses that would have avoided many costly and time-consuming dissections. Unfortunately, there was no way to circumvent the Continental Divide, and finally, quite by accident, we discovered that when the engine sputtered and stopped on the passes, if we simply climbed out of the car to stretch our legs and admire the view for about half an hour, the Jordan was ready to go again.

Our route east lay through Bakersfield, then up to Salt Lake City to join the Lincoln Highway. It was on a hot Saturday evening in early September that we pulled into the little town of Las Vegas. The streets were swarming with construction workers from the Hoover Dam site, which was just at the earthmoving stage. We were planning to eat and spend the night there, but every parking place was taken, especially around the bank, where the cars were headed into the curb to provide maximum parking. We noticed a large number of armed deputies in white sleeveless shirts but thought little about it at the time. Around and around the block we went and finally spotted a newly vacated space right in front of the bank. For fear of losing it if we went around again, we backed in and went off in search of a restaurant. The Vegas of 1932 had no neon lights or casinos; it was simply a tough, hot construction town on a Saturday night.

When we returned from supper, we found the car surrounded by deputies, whom we thought were admiring our rare and beautiful machine that looked as though it could outrun the wind. But as we approached we were seized and pushed up against the car. "What were you doing parked in front of the bank, headed out, on payday, when the money was being issued from inside the building?" We explained about being afraid of losing the parking place, that we were college boys (very suspicious), and that the rifles and shotguns were no different from those many people carried and still do in the rural West. We showed our identifications, tried to explain the car, and finally were permitted to leave if we got out of town. Tired as we were, we started the long run to Salt Lake City with a sheriff's car on our tail until we had left that godforsaken hole fifty miles behind. The rest of the trip was uneventful, barring vapor locks, as we proceeded with increasing dread and anticipation, to New Haven and a new chapter in our lives.

The Yale campus of 1932, although hemmed in by a typical urban industrial environment, maintained some semblance of greenery, and the famous elms still lent dignity and shade to the old campus where we, as freshmen, were housed in Welsh Hall. Bob Savage and I soon discovered the open spaces of the golf course (although I did not play golf) and the adjacent forested areas with their wetlands rich in pond life, rock outcroppings, fields, and sheltered forest openings. Both East Rock and West Rock also afforded a measure of wilderness and rugged beauty. Soon I became accustomed to the clanking and screeching of streetcars and the other city sounds, although it would be untrue to say that I did not miss the West almost daily. Occasionally when I wanted to indulge my nostalgia, I asked that a small piece of sagebrush be enclosed in a letter from home. When I opened the envelope, the pungent odor wafted through the stale and musty air of our rooms and provoked an almost unendurable wave of homesickness. I am not ordinarily a masochist, but there is sometimes a therapeutic value in scratching the itch.

The university was in a stage of transition and also deeply affected by the Great Depression. Applications for admission were not as competitive as in the better days to come, and the total enrollment of the classes was small as compared to postwar times. The College system based on the "Houses" at Harvard and the "Colleges" at Oxford and Cambridge in England was inaugurated in the fall of 1933. We made our choices of Colleges during our freshman year before they had had time to develop their individual characteristics. Bob and I chose Calhoun mostly because of its closeness to our classes.

These Colleges were well conceived, housing as they

did about two hundred and fifty students representing the three upper classes, as well as some teachers from varied fields. A distinguished scholar headed each college—in the case of Calhoun, Arnold Whitridge, great-grandson of the English schoolmaster and poet Matthew Arnold. In addition to the resident professors there were visiting scholars. We had two delightful and contrasting visitors: Cole Porter, who regaled us with songs and stories; and Stephen Vincent Benet who read to us in a lisping but beautiful voice excerpts from *John Brown's Body* and *The Devil and Daniel Webster,* or talked of history, particularly the mounting troubles in Europe.

At this time, the student body at Yale consisted chiefly of white, Eastern establishment private school men. We westerners were in a distinct minority and felt it. There were a few public high school graduates, some sons of foreign potentates, and a smaller group of the very wealthy including some with polo ponies. But most of us were well aware of the economy and our funds were limited. Food either in Commons or later in our Colleges was cheap and edible. For a change one could get a breakfast of two fried eggs, two rashers of bacon, toast, and coffee for twenty-five cents, leaving seventy-five cents to cover a slim lunch but passable supper. Most of us had some extra cash for entertainment and a reserve for the occasional big date or special excursion. Cars were not permitted but our community Jordan found an inconspicuous home at the edge of town giving us a wider scope of activities than some.

There were few "joiners" in our Western contingent. I boxed a little, hiked a lot in the forest preserves, and rode when an ROTC horse was available. Otherwise my freshman year was mostly work and a fair amount of longing for the California or Northwest mountains.

Although my preoccupation with natural history and outdoor activities had been avid and continuous, most of my formal education had been nonscientific and my knowledge of wildlife self-taught by voluminous reading and the experience I had gained by observation in wood and field. Now I had an opportunity to take courses that equipped me with some basic knowledge of biology, zoology, and botany. However, given the economic depression (President Roosevelt ordered the banks closed shortly after we had arrived in New Haven), it seemed wise not to overcommit to one field. Accordingly, I added courses in mechanical drawing and economics as insurance in case my assumed life's work in some field of the natural sciences did not guarantee me a living.

In my sophomore year I had to declare a major. During the rest of my college years I relegated the science courses to a secondary place and concentrated in a newly formed department of study entitled, History of Arts and Letters. This department had gathered a group of the greatest teachers I have had the privilege of knowing: Chauncey Tinker lecturing on the nineteenth century poets with such absorption that he limped across the platform when discussing Lord Byron; or Johnny Berdan, sprawled outrageously in his small office with a stack of the daily themes I had produced, sometimes on those sterile evenings when inspiration there was not. He thumbed through those pages, occasionally tearing one up and dropping it into the wastebasket, criticizing another in the most caustic terms, and if I were lucky, giving a precious word of praise to the short passage that showed some promise. He could spot the phony, the overblown, the "fine writing," and I soon learned to omit subjects and passages that might draw his ridicule. To this day I feel a guilty twinge when I overwrite.

During those years my interest in history grew with each new excursion. The Far East was revealed, the ancient world, Europe, but most fascinating of all, English history. Civilizations were presented through the eyes of contemporary statesmen, the works of contemporary artists, the words of poets and writers of prose. We were required to write term papers on subjects of our choice, and I chose to write about such varied topics as Max Beerbohm, the Whistler-Ruskin

Trial, the Jardin des Plantes.

Only recently I learned from a professor at Yale who had been involved in this department that the History of Arts and Letters program was discontinued—at least under that title. I wish now that I had asked whether this was from lack of student interest or an administrative judgment that it lacked relevance. In either case, I am sure that it is a serious loss to students at a time when history is largely ignored and relevance is judged chiefly on the basis of what immediate effect an event or decision will have on the individual. Judgments or decisions about problems of today are frequently uninformed—effects where causes are unknown, a shell game in which there is no pea of knowledge. I became steeped in the thoughts and deeds of other people in other times, the poetry I learned turned the rhythm of tires on a concrete pavement into iambic pentameter, and knowledge of the wise acts and decisions of courageous men and the follies of Light Brigades all provided perspective with contemporary application.

For diversions there were the occasional carefree weekends at one of the neighboring women's colleges and visits to friends at Harvard (in fact I spent so much time there during reading periods that I even held a meal ticket at Elliott Hall). Zoology and botany fieldtrips and outings in the Jordan provided the outdoor activities I craved. I had joined ROTC, not so much out of patriotism, as for the field artillery battalion with horse-drawn French seventy-fives and caissons that were still maintained at Yale. It was great fun to ride a wheelhorse and go slamming along with gun and caisson bouncing behind, especially as we were permitted to "practice" riding between regular sessions. One year some of us who were experienced horsemen agreed to participate in the ROTC event promoted by the New Haven horse show. We made a pact that if any of us won the fine English saddle that was the prize, he was to sell it and throw a party for the group. I won the saddle and sold it. We had a party with the best of gin we could make (one gallon distilled water, one gallon grain alcohol, fourteen drops of juniper extract, and two ounces of glycerine to smooth it all out), and lobster dinner at Savin Rock, all paid for by the proceeds of the sale to the local saddle shop.

Coming of Age in Mexico

By spring my roommate and I were already thinking about plans for summer. One of the students, Bill Prestley, who shared the living quarters with Bob and me, came from Mexico City where his father had retired after many years as a physician with a mining company. He suggested that if three of us chipped in and bought an inexpensive car (the Jordan was unsuitable), we would all save money going home by way of Mexico City, where we could visit him for the term of our entry permits. This rationalization sounded fine, and we were able to buy a 1929 Ford touring car for three hundred and fifty dollars. After obtaining permits but neglecting to inquire about road condition, Bob Savage, Bill Prestley, and I packed up our camping gear, spare gas cans, lots of tire patches, water bags, and started off for Laredo, Texas, where the new international highway was purported to begin.

It was very hot in the open car, and we were eager to cross the border, but first we stocked up with cartons of Chesterfield cigarettes, filled all gas cans, and provisioned. At the Laredo gate the customs officer asked us our destination. When we told him we were planning to drive to Mexico City, he said the road was not complete and denied us permission to enter the country. It was a classic impasse.

We were so politically naïve that we were unaware of conditions in the Mexico of 1933. Had we been better informed, we would never have undertaken the trip, thereby missing a fine adventure. We concluded that the problem lay solely with a disagreeable border guard, and as the afternoon was passing into evening, we first ate supper, and then, assuming the relief shift would be manning the gate, we tried again. This time we stated we were heading for Monterrey, Mexico, which was quite reasonable, and a bored officer who never even asked the purpose of our visit to Mexico passed us through.

The uncertain and chaotic situation that existed in Mexico that summer of 1933 began after President Alvaro Obregón's enlightened regime had ended in 1924. He was followed by Plutarco Calles, who after his two-year term as president, remained the power behind the throne until 1934 when Lázaro Cárdenas succeeded and restored a more just and orderly regime. During Calles's domination, however, a one-party government was created (the National Revolutionary Party). There were bitter conflicts over mineral rights with foreign firms. Anticlericalism was rife, and all church lands were expropriated while militant Catholic groups broke into open revolt. They called themselves Cristeros, and the fierce clashes resulting in extremely harsh repression by Calles's troop continued throughout this time of radicalism, fanaticism, and corruption.

We were blithely unaware of all of this as we rolled along the blacktop road through moonlit sagebrush and desert country trying, sometimes in vain, to avoid the desert tortoises that abounded along this desolate stretch and seemed to enjoy the smooth, warm surface of the road. We anticipated some rough highway but so far conditions were favorable. Before midnight, we drove off the road and out of sight behind a little patch of scrub, rolled out our sleeping bags, and slept soundly. We awoke with the sun, cooked a quick breakfast, and started off again, estimating that we were about six hundred miles from Mexico City. The road continued to be fairly good as far as Monterrey, where we gassed up, bought fresh food, and started off again.

It was about two o'clock that afternoon when we met our first obstacle—a ragged and knotted rope stretched loosely across the pitted road and tied at both ends to insecurely propped posts. A dusty path led from the post on the right to the top of the low bank where a makeshift, open shelter of corrugated iron with an adobe cooking platform stood. A few skinny molting hens scratched in the dirt and a handsome gamecock was tethered to one of the roof supports. A fat, dark woman with a broad Indian face looked down at our car from her post in front of a tiny fire, while a stout, barefooted man wearing a worn army cap and a filthy military jacket that barely met his old brown cotton trousers, lounged in front of the shelter on a broken-down deck chair. He looked surprised when he saw us, for certainly he was not expecting much automobile traffic on this road. He got up slowly, shuffled into his huaraches, made a sketchy attempt at straightening his semiuniform, and half sliding down the bank, approached our car and leaned on the door of the driver's side.

Bill Prestley politely asked what was going on. The man replied that this was an inspection stop and began to search through our gear piled in the back. When he found the cartons of cigarettes, he removed two of them, slowly untied the rope, and was about to wave us on when a Mexican peasant came up behind us driving a burro colt, a few goats, and two brown sheep. The rope was hastily raised again while the soldier-bandit (for that was what he was) walked casually over to the little flock, carefully felt the sides and back of one of the sheep before picking it up, calling to the woman to come and take it to the shed. Apparently the poor paisano did not dare object. The rope was again lowered and we all passed through.

We felt anger and consternation, not realizing that the countryside in this area was scattered with similar little groups of soldiers who settled down wherever they happened to be when the latest unrest had been quelled and pay had been discontinued. They usually found some woman to share this simple life of easy plunder that called for little exertion. We wondered why all our cigarettes and other gear had not been taken. About fifty miles farther along our way we found our answer when we were stopped by another barrier, this time controlled by a group of about ten soldiers and camp followers. The routine was the same: we lost more cigarettes, some canned goods, and our largest cooking pot. There is honor among thieves, for like the first bandit, these, too, left some loot, which warned us that there were to be more inspections ahead. The next and last "inspector" generously left us with a few packs of cigarettes but took socks and some other items of clothing.

The next day we reached Ciudad Victoria, still on fairly decent dirt roads, but then the road itself virtually ceased. During the entire journey, we had seen no road-building equipment or any construction activity, and now it seemed the whole project had been abandoned. The roadbed was ballasted with four- to six-inch sharp rocks and deep transverse ditches for eventual culverts and could be circumvented only by finding a way to the desert floor, around the ditch, and back up to the roadbed. Soon we began to climb, gently at first as the desert gave way to a scrubby jungle and detouring became far more difficult. We began to meet little groups of white-clad Indians on the road and saw primitive shelters nearby, as obviously this cleared ribbon through the brush was replacing the older trails. The people were small, delicately made, and dressed in immaculate white sheetlike robes. Surprised and alarmed by our appearance and by our car, they scuttled off the road driving their goats and burros, then stood and watched through obsidian eyes in Aztec faces that might have come to life from a temple carving. Apparently they neither spoke nor understood Spanish.

It soon became obvious that we must drive long hours and into the night if we were ever to reach Mexico City before our diminished supplies were exhausted. We were grateful that the bandits had no use for gasoline or we would have been stranded long ago. We were averaging about ten miles an hour when the punctures began to occur from the needle-sharp pieces of flint and volcanic glass that were present among the larger rocks of the road ballast. We hardly noticed the first flat because of the constant jouncing, but fortunately we stopped before the tire was ruined. We confidently dug down under the front seat for the jack, but found none. Our careful planning had a flaw in it. Fulcrums were easy to find, but levers of sufficient length and sturdiness were scarce in the scrubby jungle country. Luckily our ax, under the front seat where our jack should have been, had not been found and "liberated," and we chopped a fairly straight pole to use as a lever. Then after unloading most of our heavy gear, we were able to raise and block up the axle. We had two spare tires, one mounted on a wheel, and from then on we developed great teamwork: unlash the pole tied to the side of the car, raise, block, pry off tire, patch tube, tramp tire back over the rim, inflate with hand pump. I think we got it down to about twenty minutes. I do not know how many tubes we repaired, but we had enough patches to take us through.

When driving at night, one of us sat on a front fender, grasping a strap we had attached to the hood and looking for hidden culverts, resting livestock, or super boulders through the dust that swirled around us. We drove until we were seeing prairie schooners, giants, and solid walls in front of us. Then, leaving the car in the middle of the road, we moved into the brush, made a fire, cooked a simple supper, and curled up in our sleeping bags.

Twice we woke in the morning to find ourselves ringed by white figures, gazing down at us, expressionless, each carrying a formidable, brush-cutting machete and a gourd hung from the shoulder. Few things make one feel more helpless than to be pinioned in a narrow sleeping bag. One discovers that one is capable of a whole repertoire of conciliatory cooing sounds, jolly laughter, pleasant smiles, all of which bounced off inscrutability without a trace. "Buenos días. ¿Cómo está usted?..." No comprehension evidenced.

We eased ourselves to a sitting position and tentatively extended a pack of cigarettes. No reponse at first. Then one of the small figures stepped forward and offered his gourd to one of us, who, not wishing to offend, gulped a great mouthful of liquid fire, sputtered, and turned red. This broke the tension. There were smiles of a sort. We climbed out of our bags, offering to show them the car with graceful, bowing gestures, but they looked suspicious, and quite suddenly turned away, climbed the bank to the roadbed, and disappeared around a bend.

By now we were high in the Sierra Madre Oriental, having started the steep climb after obtaining supplies in the small town of Tamazunchale. The road became increasingly hazardous as it wound its way along narrow cuts in the mountainside without benefit of guardrails or shoulders, only to descend into valleys where the streams were usually shallow enough to ford. On two occasions crude log rafts were necessary to ferry livestock and natives across deeper, wider waters, and we were permitted to jounce our car aboard and cross where a future bridge was planned. We paid in pesos and received restrained, but not unfriendly assistance.

This mountain country was magnificent—cool at the higher altitudes and lush where there was moisture. I would enjoy driving that road effortlessly now to bring back the memory of those ten days without distractions. For at that time we were so occupied with watching the road ahead, patching tires, and wondering if we would ever get through the great mountain range, that we missed most of the natural beauty and much of the color of the scattered settlements and the age-old seminomadic way of life. We glimpsed only briefly the small shelters and traveling bands of immaculate white-draped figures with narrow, slanted foreheads merging into aquiline noses—and always the expressionless black eyes.

On the ninth day we emerged from the mountains into the high basin plateau where ancient Mexico City lay. We saw the water gardens glittering in the distance and the pyramids and temples of the Aztec empire. Twentieth century villas crouched beside cathedrals, monasteries, and ancient ruins. Commercial buildings crowded the base of the central hill where the nineteenth century palace of the ill-fated Carlotta and Maximilian stood.

Industrial development was scarce in 1933 compared to today, and the yellow pall of smog, now so much a characteristic of the city, was absent. The great, snow-topped mountains with the names impossible for the English tongue to scale—Popcatépetl, Citlaltépetl, and Ixtaxcihuatl—stood above the rim of lesser mountains against the clear sky.

We were greeted rather coolly by Doctor Prestley and his matter-of-fact wife. They were apparently expecting us to come by train as they knew that the highway was not due to be completed for another eight to ten years, and they had anticipated a telegram giving the date of arrival. When we described our journey, they were really quite angry with their son, who after all should have known conditions in Mexico; however, their relief that all was well soon melted the ice and a most varied and pleasant visit began with cold cervesa and baths.

We explored the great city from its seamiest depths (and there can be few seamier) to its most glittering examples of "modern architecture," from the great cathedral, the Zócalo, all built upon or abutting ancient walls, to the temples and truncated Aztec pyramids—a city of contrast. We also went to cockfights, bullfights, and one day took off with a full tank of gas and a twenty dollar bill for a short drive south of the city.

We kept on driving, up over the passes where the air was fresh and thin, then down into the broken, arid brushland, surprisingly green and fertile in those mountain valleys where there was water. About sixty miles south of the city we rounded a bend of the poorly surfaced hillside road, and there, in the midst of this dry, forbidding landscape, was Taxco—pretourist Taxco. White-walled houses climbed the mountainside. The steep cobbled streets were too narrow for cars, except for the one that formed a continuation of the road leading into the tree-shaded plaza that was dominated by a marvelous, two-towered Spanish colonial cathedral. Some streets were decorated in Indian designs formed by a crude mosaic of darker stones. It was already evening when we arrived and Saturday at that. The promenade had started—girls in one direction, boys in the other, around and around the plaza, chins high, pretending to ignore one another, occasionally exchanging a look, a smile, in the age-old ritual. Blackbirds going to roost screamed in the trees overhead, and goat-tailed grackles stalked among the strollers unconcernedly.

Gradually the sun sank, red-orange in the west. The white sheets shading open market stalls, the red-roofed, white-walled buildings, the looming presence of the pink cathedral from whose dark interior gold-encrusted carvings glimmered in the light from the high windows, all created an overpowering harmony, a patina of history and cultures, where edges blurred and shapes of man and forms of nature blended.

Dark faces looked at us curiously but without hostility or calculation, for tourists in numbers had not yet invaded this enchanted hill town. Even today with the rich silver mine and the tallers, the new hotels, the many visitors, Taxco retains much of its original wonder, clinging and shining on the steep mountainside and protected as a national treasure from overexploitation.

We were enthralled and wanted to stay on, but since we had only the American twenty dollar note that no one would cash, we decided to take a chance. We sent the bill with a message to our Mexico City hosts, telling them that we were not coming home for a few days and not to worry. The sealed envelope was entrusted to the driver of the rickety bus that formed the only link between the city and Taxco. To our surprise, he returned the next day with many pesos, making it possible for us to stay on in the little three-bedroom hotel, and eat our "hot" meals in one of the two cantinas. We rented three skinny, small horses and rode down into the valley where old walls and mine structures were being reclaimed by the twisting buttress roots of trees and vines. A stream coursed through the valley, and as we approached it to water our horses, several iguanas, lying camouflaged on overhanging branches, launched themselves into the pool and swam smoothly to the shelter of a crumbling dry wall. We also saw a coati, tail erect, hunting along the bank and poking his long, flexible nose into crannies while, coonlike, he turned over small stones and rotted timbers.

On the third night of our stay, there was a murder in the usually quiet little town. Apparently the uncle of one of the young men had tried drunkenly to make love to his nephew's bride in the cantina on the square. She protested, then cried for help, and in front of many of their friends, the boy stabbed his uncle in the chest and he bled to death in a very short time. An uproar resulted and sides were taken as to whether the whole matter should be hushed up or whether the police from Mexico City should be involved. I think that our presence made secrecy seem impractical, and we suddenly became the objects of suspicious looks and discussion. We felt acutely uncomfortable. It seemed wise not to become involved, and so we strolled from the scene, then quickly climbed into our car and headed back to the city. I believe the police were notified and the boy taken to trial in the capital—a sad ending to a young marriage.

Our visiting permits were only for a month, and what with the unexpected difficulties of our trip and the excitement of each day, we left renewal until the end. We had already decided that we would take the train rather than drive back, especially as the rainy season had started. Accordingly, we presented ourselves to the appropriate government office at 11:00 A.M. and sat waiting on a long, wooden bench against a wall facing a desk where a chubby girl was sitting. She stated that the official concerned with such matters was not in that day and that his secretary was also not expected.

We went home and reported to Mrs. Prestley, who was not surprised, but stated that this called for a little present-giving and a more oblique approach. She accompanied us the following day, and with a mixture of scathing comments, interspersed with expressions of gratitude and obligation should our request be granted, was promised that if we returned in two days we could pick up the extensions. Five days later we became legal visitors again.

Bob and I bought train tickets (Bill Prestley was not returning with us) and arranged to ship the car ahead by rail freight to be off-loaded in Salteo, from which point a good road led back to Laredo.

As a going-away present, the doctor presented each of us with a quart of the finest Scotch whiskey—a straight malt, especially prepared for the British Club of Mexico City. We divided these two quarts into eight one-half-pint bottles, and cutting a slit just long enough, we slid these smaller bottles into an inner tube, with tissue paper in between. We were able to fit them all in, repairing the tube well enough to hold air. We then mounted both tube and tire on our one spare wheel and off went the car. We felt serious doubts as to whether we would ever see it again. Three days later we climbed aboard the comfortable and well-appointed train and started our journey north.

After settling our luggage in our reserved compartment, we walked back to the club car where some of our fellow passengers had gathered. Soon a handsome, gray-haired man with a quiet, assured manner sat down beside me, smiling in a friendly way. He was a General Motors representative who had lived in Mexico City for many years and took this train routinely. He had long ago given up trying to make a Mexican factory run in the manner of its American counterpart, or convert a Mexican into an efficient factory worker, and had accepted with resignation the spirit of his adopted land. He said he both loved and hated Mexico with its filth and slovenliness, its graft and brutality, its color and its warmth. His words were at one moment poetic, at the next expressed disgust, but one felt that he had a profound understanding, not judgmental but accepting of the old patterns as they had always existed.

Slowly the evening dragged on, and the passengers went off to their respective berths until only the General Motors representative, Bob Savage, and I remained. We ordered highballs and sat smoking quietly, when our new friend broke the silence. "Strange what people come to Mexico and why. Every kind comes at one time or another seeking something they cannot find at home. Many of them don't know what it is—they just want a change. That young Englishman—he's probably spending a small inheritance in the modern version of the grand tour. There are lots like him. They go everywhere, walk about with their eyes open, and impose on everything their own interpretation, their own prejudices. He will go back to England with a precise picture of Mexico. He might even write a book about it—something like, 'What is wrong with Mexico?' and people will read it and say, 'Very interesting,' and then forget it all.

"What's wrong with Mexico? Everything is wrong with it, but he won't understand why or how hopeless it is to try to do anything to change it. It's the hard soil, the hot sky, the ancient dead glories, and the old corrupt church which holds a mystic power and finds itself and its tenets all inextricably mingled with the primitive cults and superstitions. It is the hopelessness of the whole structure that people fail to understand. Have you read D. H. Lawrence's *The Plumed Serpent*? It's fantastic. But he understood better than most the dignity that comes from the very attempt to progress where knowledge of its futility is at the very heart of Mexican consciousness.

"What's wrong with Mexico? It isn't its economic backwardness, it isn't its newly discovered raw materials, but it's the spirit of the land."

I interrupted this monologue, asking, "But if this is true, where do the fighting, the violent social upheavals, the friendly smiles come from? Are these signs of this hopelessness?" "Yes." He paused a moment. "But it's no use explaining it. It has to be seen and felt. All this is an individual expression of momentary revolt. The smile is a personal reaction to the hopelessness. It is valued as a fleeting pleasure, and it is not followed by the expectation that the next minute there will be occasion for laughter.

"But let's not talk about it anymore. I sometimes think so much about it that I can't see it in proportion myself. You two may have a better idea of what is wrong than I. Perhaps this hopelessness is the valiant admission of realities which we do not appreciate. I cannot tell, not being quite one of them, and they can't know because it is their avatar.

Now, many visits later I think of the club car conversation and wonder if the appearances of Mexican modernity are but a shroud concealing dichotomies still as alive in this fascinating land as when I jotted down these notes more than fifty years ago.

It was about noon on the following day as the train skirted a hill along a shallow cut above the barren desert when suddenly there was a fearful jar. Air brakes screamed, the train plunged backward, rocked, and then there was silence. The passengers looked at one another in dazed fright, then started for the doors. Ahead, just around the bend of the hill, the tracks rose, twisted in fantastic shapes. Splintered ties still clung to some sections. A steaming locomotive lay against the bank with its cowcatcher and wheels sunk in the torn roadbed. Behind it lay broken stock cars, some having rolled over the bank into the valley fifty feet below, while only the caboose still stood upright on the rails. It took a few minutes to comprehend the situation. Gradually the pungent smell of burned flesh and drying blood seemed to roll forward and encompass us in nauseating waves. Then we heard the sounds—the agonized, almost human cries of horses, mules, and cattle imprisoned in the cars. They were frantic from injuries inflicted by splintering boards, from the heat, and the hordes of flies that formed black rings about their eyes and mouths and wounds. The only men at first visible were a few soldiers, standing quite stolidly by the engine.

Something had to be done and no one was doing it, so Bob and I ran forward along the tracks, almost stumbling over the body of a big Mexican who lay on his side facing the passenger train. His head was horribly cut and one leg was so swollen that it seemed almost to burst his faded jeans. At the knee a tear revealed the kneecap lifted. Part of the cloth as well as bits of coal and wood were thrust into the wound. He raised his face to us and gasped unintelligibly, then his head fell back on his arm. In one hand he still clutched the flag with which he had stopped our train. He must have dragged himself the hundred yards or so in his brave effort to save us. It was not until afterward that we wondered why none of the soldiers who must have been riding in the caboose had not saved him this agonizing journey or offered help. We lifted him to a camp cot that someone dug up from the caboose. While one of us started to clean the wound, the other went to search the wreckage for a first aid kit and release the frantic animals. Three more men were found. The fireman, crumpled over a coal shovel, was dead. His skin clung to his clothes when he was moved, and the smell of scalded flesh filled the cab. The engineer lay half-conscious beneath a pile of debris. His chest was crushed and his eyes bulged as in agony. Gradually the heat of the day subsided, and the animals that could move were grazing in the valley below while many cars were still partly filled with the carcasses of severely injured beasts that we had killed by poleaxing them with the fire axes. After much persuasion, the soldiers helped carry the human bodies and the wounded men to the caboose of the freight train, and feeling that we had done all we could, we returned to the club car.

The conductor, all graciousness and smiles, thanked everyone and assured us that telephone contact would be established via the wires on the pole line accompanying the track. Soon a train would be backed down from Salteo some sixty miles away and we would be assisted around the wreckage to continue our journey in comfort. Meanwhile, everything was to be on the house.

We all took advantage of this generosity. The general shock and horror of the occasion was soon drowned in noisy talk, forced laughter, and abundant drink.

But help did not come soon, although a crew of workmen appeared. They lit flares along the path of the wreckage as darkness fell and began clearing some of the wreckage without much apparent plan or supervision.

By midnight the liquor had run out, and, more importantly, the water supply was exhausted. We were issued some small bottles of mineral water for washing purposes as we straggled off toward our berths. Bob and I were sobered, however, by a comment from our friend from General Motors. He tactfully expressed sympathy with our energetic endeavors in caring for the wounded men and the killing of dying animals, while freeing others by breaking into the stock cars. He warned us, however, that the authorities in Salteo might react strongly. Either we would be graciously thanked or accused of unauthorized interference and destruction of property, or worse. Neither of us slept well, thinking of provincial Mexican jails and other horrors. The next morning, a thirsty, hung over group of passengers was assembled in the club car and informed that we must gather up our baggage and scramble down the embankment around the wreckage to where a rescue train was waiting. They were sorry that they could not help much. Only the train crews were available, and they were not porters, so we must do our best. Accordingly, the ragged procession started winding its way along the desert floor, while Bob and I watched from above.

About two hours later we pulled into Salteo, where a large group of officials were awaiting our arrival. Bob and I hung back as long as possible, trying to judge the mood, but finally our General Motors friend, while disassociating himself from us as fellow U.S. citizens, advised us to come forward. He said he thought all was well, and he was right, for when we emerged, it was all smiles. We were thanked officially, given warm "abrazos" with vigorous back-patting, and invited to an official ceremony of appreciation. We felt that as time passed and more of the story unfolded, the mood might change. With many thanks, we declined the banquet, and to our delight, were escorted to where our car stood by a siding.

The battered Ford started right away, and we were off to the border at our best speed, reaching Laredo and crossing into the blessed land with no serious customs inspection. We parked the car for servicing in the garage of the best hotel in town and ordered a meal of steak, French fries, milk, and salad, then returned to the car after dinner to retrieve our toilet articles.

The garage was fully lighted, and a delicious odor of fine malt Scotch whiskey filled the air. Our car had just been backed off the rack when an amber pool began to spread across the concrete floor from the rim of a flat rear tire. The shock on the face of the mechanic was comical as he explained that he had noticed that one of our tires seemed to be going flat and that he had decided to put on the spare while he had the car still on the rack. As he had started to back off, he had heard the sound of crunching glass and stopped at once.

He was more than willing to help us salvage what we could. This time we jacked up the car without moving it, pried off the tire gently, and carefully extracted the tube. Miraculously, two one-half-pint bottles of Scotch that were at the top of the wheel were unbroken and saved for a later celebration. The garage floor was hosed down and the tube and broken glass disposed of discreetly. That night we slept the sleep of the righteous in American beds and in our own sweet land.

Head Over Heels at Yale 

On coming back to Santa Barbara after the Mexico trip, I found myself restless and unwilling or unable to fall back into old patterns. I suppose the year away in New Haven had brought about some maturation, but also my friends had rearranged their relationships, some finding jobs or new interests in which I did not participate. Previous romantic alliances had altered and recombined, while I had no active attachments at the moment. Fall, with the return to college, was only a month away and I was at loose ends.

After a few days at home, I decided to call a girl who had just completed her freshman year at Sarah Lawrence in Bronxville, New York. Connie Hopkins and her family had come to Santa Barbara a year before from Evanston, Illinois. She had spent the two and one-half years prior to that at various schools in England, Paris, and Switzerland. Although Connie and I went around with different crowds, we did take notice of each other at the beaches, and I was acquainted with a number of her classmates. I knew Connie was musical, studied piano, and that she and her two brothers flew those small open planes of the day, looking dashing in black leather helmets and trailing scarves. I had invited her to the freshman prom, but she declined most sweetly because, she said, she already had an engagement. (She now claims that she refused, because she was afraid I might be disappointed in her.) I did not go to the prom, but it was then that I decided on a return engagement.

I called Connie, asked her to dinner, and she accepted. That evening was magical. She wore a dress of dark color with a brocade top and bouffant sleeves. Her long black hair, growing thick in a widow's peak was done in a low knot over her slender neck, and her Celtic blue eyes and white skin were cameolike in their delicacy—all this, together with a five-foot-two slender form and sweet open expression! One cannot pick one's parents or one's children, but one certainly should choose one's wife with great care and deliberation. The impact she had on me made this deliberate and careful selection a simple and swift one. By the end of the evening, the dancing, and above all, the talking, we found that we knew many of the same places, that even her school in Lausanne was within walking distance of Les Abeilles. Our differences seemed only to lend enchantment. By the time the door of her family's beautiful, Spanish-style house had closed on a good-night kiss, the hint of dawn was rising

Flying a small open plane was one of Connie Hopkins' hobbies in 1933.

above the mountains to the east. I was sure that I had found my future.

She knew little about the outdoors and the wildlife interests that were so central to my life, while I knew very little about the music and mysticism that had been part of her upbringing. We looked forward to enlightening each other.

It was hard to return to New Haven leaving Connie behind. She and her parents had decided that since she was a serious music student she should give up college in the East, move to San Francisco where she found a studio apartment in San Francisco's Chinatown, and study piano in earnest with a woman teacher of some repute. She was only eighteen, lonely much of the time, and soon realized that although she was gifted she would never attain real virtuosity. With plans for our marriage after I graduated, the career of a professional musician did not seem so appealing. But she tried to satisfy her own and parental expectations. All the while, as we wrote and planned, the endless hours at the piano seemed less in context with the future we were envisioning. She continued for two years, and after practice hours worked with settlement children in musical projects, which she thoroughly enjoyed. We wrote endless letters, for one did not telephone cross-country in those days. For one Christmas and for both the following two summer vacations, she interrupted her work and came to Santa Barbara so that we could be together. Each occasion reconfirmed our commitment, which during the long months of separation would slide a little for one or the other of us.

My mother and Tom welcomed Connie with delight and relief. ("Think of what he might have chosen in view of his past record of susceptibility and fickleness.") Dudley Carpenter approved with a nod and his wise, understanding smile, which helped gain Connie full acceptance in our teatime circle. And above all, she won her own way with all our friends because of her charm and beauty.

During those summers I introduced her to my world of

camping and hiking, and she, in turn, played her music for me. We discussed at length our sometimes differing beliefs and philosophies but came together in our mutual love of books. Fortunately her sister, and more cautiously her protective brothers, accepted me and we became friends. Understandably, her parents were more skeptical and anxious, for I fitted no model of stability and showed no solid evidence of an ability or even intention if we married of maintaining their daughter in the manner to which she was accustomed.

Then intervention came in the form of Connie's grandmother—a step-grandmother, really, but in a true sense the only grandmother Connie had known. She was a dignified, fine-looking upright woman whose early life as a down-Maine schoolteacher had hardly prepared her for a late, childless marriage to Connie's maternal grandfather, Charles Stinchfield. She undertook the job of governing a large, brilliant, and erratic family of "steps" and "halves" with common sense, fortitude, and understanding, accepting with

Connie and David on horseback, 1935.

grace the wealth, position, and early widowhood that accompanied it. The tenets of her Spartan, New England background remained uncompromised.

She was a keen observer who loved Connie and seemed to like me. She realized that our engagement, now formally announced, was placed under needless strain by further continentwide separation. She also understood that the lonely life and intense musical training that had kept Connie faithfully working except for the two summer vacations, should end. Accordingly, she suggested quite firmly that Connie go to Boston during my senior year, where she would attend the Garland School of Homemaking, which taught the fine arts of keeping a husband happy, well fed, and proud of a wife who could not only command all the social graces but sew a fine seam. Of course, she was actually saying, "For goodness sakes, let's recognize that these two seem determined to marry, have been patient, and the time has come to give them a chance to be together before they run off and do something silly."

And so it happened. Connie was installed in a brownstone house on Commonwealth Avenue with a bevy of nine other attractive and varied types of girls, all biding their time before marriage and humorous about the way they were doing it. For me the whole situation was great. By keeping my grades high enough to stay on the Dean's list, I had unlimited weekends off, and with the car it was not far to Boston where I scrounged quarters and meals with my friends at Harvard. I became accepted at Garland as "the fiancé in residence" and could count on experimental meals at almost any time while curfews were loosely enforced.

Perhaps I should explain my financial status, lest it appear that I was rolling in money and could indulge any fantasy. When my father died, he left trust funds to my brother and to me that were to be administered by my mother. When I went to college, she turned over the income, which had shrunk drastically during the Depression. It amounted to about $175 a month, not inconsiderable in

1930s dollars, but still, after college tuition, board, clothes, travel, and so forth, left a rather small residue for books, entertainment, and weekends. Weeks were largely spent in a monastic, scholarly atmosphere, with little distraction, but, ah! —the weekends—one planned for those. There were the football weekends, with or without a female guest. There was Derby Day, a dance at a woman's college, or a weekend in New York. That was the best and most expensive indulgence.

For such occasions, gray spats were not out of line. Tails and even a collapsible opera hat were part of the well-dressed, tailored by Langrock or J. Press, Yale man's wardrobe. Sometimes I bought brand-new unclaimed clothes from these prestigious institutions, which I then had refitted at a considerable saving. At school in California and during the summers I had felt most at home in Levi's, cowboy boots, cowboy shirt, and leather vest, but for the interval in New Haven, I enjoyed dressing in the style of the day.

That last year at college was a good one. Connie and I and our friends made the most of those weekends, all dressed up on a New York excursion. The girls stayed at the Barbizon and we at the Lexington Hotel. We tea-danced at the St. Regis or Waldorf and in the evening to the great bands: Benny Goodman, Tommy Dorsey, and Glenn Miller. We slithered and dipped, cheek-to-cheek, stardust all about us, smoke in our eyes, round and round night and day. With a million others we danced the beguine and knew that Ray Noble had written "Love Is the Sweetest Thing" with us specifically in mind. Prohibition was over. We no longer carried our silver flasks, made gin, and frequented the "athletic clubs" of New Haven or some of the intimate or not-so-elegant New York speakeasies of the freshman and sophomore years.

But as the senior year began to draw to a close, increasingly the specter of joblessness occupied our minds. Times were still bad in 1936, and there were no representatives of large corporations waiting to interview us. For the majority of us who had only a general liberal arts training, the prospects were indeed bleak.

I applied to the Field Museum of Natural History in Chicago for some sort of job and was tentatively offered a position in the specimen cataloguing department, but this held no future. Furthermore, both Connie and I wanted to go west, preferably to the Northwest, far enough from family ties, yet well within reach for visits. I suppose I could have sold cars or real estate in Santa Barbara, or perhaps taught in some private school, but for us the idea of settling where there were mountains and lakes, fishing, hunting, and the smell of pines and firs, and sage across the mountains—that was what we really wanted.

Tom Ripley, my stepfather, had gone to Tacoma, Washington, in 1889 and had eventually become the president of Wheeler-Osgood, a door and plywood company. He realized our dilemma and asked the trustee management of the company from which he had retired many years previously and which was now in dire financial troubles, if they would give me a job. They agreed. With no salary named and no job description, I could now claim that I had a position and that accordingly, Connie and I would marry immediately after graduation.

Her mother had been quite ill that spring and was not able to make arrangements for a formal wedding. We decided that it would be simplest for everyone, since Connie's brothers and sister were in the East and my mother and Tom were planning to return from Europe at that time, if the wedding were to take place in Dwight Memorial Chapel on Yale campus that twentieth of June, the day following my graduation.

We were to be married by the Reverend Baldwin from Baltimore, age ninety-seven, and oldest living Yale graduate, who was a great-uncle of Connie's and was to be in New Haven anyway for some ceremonial rite. He was accompanied by his spinster daughter, Mariah, who watched over him with grim, unsmiling fortitude, interrupting his garrulous excursions into the past when his audiences tired and drifted away—a desertion his near blindness did not appreciate. He was also palsied and hard of hearing, but withal a huge, rather magnificent craggy sort of man who talked of Lincoln and the "war" as though it were yesterday. When we met him the night before the wedding and attempted a rehearsal, it was evident that the ceremony might be a hazardous undertaking. We arranged to have an old friend of his in the Physics Department at the university prompt him if he wandered too far afield. Connie and I did not care or worry so long as we were married and on our way west the following day.

Graduation ceremonies passed smoothly in the fine June weather as we sat listening to words of inspiration from our mentors and self-consciously smoking our long, white clay pipes. Then it was over and Connie joined her family at the Taft Hotel, while I joined Tom and my mother at a hotel in Hartford to wait for the morrow. I had no prewedding doubts or jitters. It was warming and sentimental to be with my mother and Tom, both of whom were so close to me and happy about my choice of wife. We laughed a lot, and my mother even had a drink or two. The long evening was interrupted by the broadcast of the first Schmeling-Louis fight, which caused much excitement in sports circles. I had wagered quite heavily and at good odds that Schmeling would win. When the broadcast started, one could feel through static the excitement of the great crowd. I strained to hear the blow-by-blow account while my mother expressed horror at the brutality and recoiled at the sound of each thump of glove on vulnerable flesh. As the clamor mounted, I turned down the volume to try to hear the description more clearly, but my mother, shuddering, said "It's bestial! Would you turn that up a little bit?" And we all three burst into laughter. Schmeling did win that one, and my winnings helped in our journey west.

The next day our wedding party was all ready on time at the chapel door but had to wait outside while a damp drizzle started and a preceding ceremony dragged on. Finally it was our turn. I do not remember much about the wedding except that Uncle Charles Baldwin immediately and irretrievably lost his way in the ceremony and defied all attempts by his friend from the Physics Department to rescue him. Connie's father stated later that we had been half-married and half-buried, but be that as it may, we had a certificate. We all adjourned to the ballroom of the Taft Hotel for a champagne reception.

Uncle Charles and Aunt Mariah were escorted to a place of honor. The reverend instantly realized that even if he could not see clearly he had a large potential audience and he was off. Unfortunately, both he and Mariah were uncompromising teetotalers and in consideration of this, arrangements had been made for soft drinks, tea, and coffee to be served first. Then, in the expectation that the old man would be tired out and retire after a few minutes, the champagne would appear. Uncle Charles never tired, nor did Aunt Mariah, and finally the guests left and the champagne, unopened, was given to the hotel staff to dispose of as they wished while the wedding party went off for lobster dinner and jollity at Saven Rock.

Connie and I were all packed and ready to go. We had bought a Ford convertible of some age but in good condition, and all our clothes for the trip were laid out ahead of time. Hartford was our destination as it was late when the party was over. The next morning we picked up our dog, Linda, at a kennel in Needham, Massachusetts. She was a beautiful Norwegian elkhound that we bought as a wedding present to each other. At that time elkhounds were rare in this country. I had seen pictures in the papers of two of them with President Hoover in the White House, and their solid conformation, alert expression, and balanced symmetry appealed to me as had no other breed. So our family was increased to three as we started in high spirits down the Mohawk Trail.

I will only mention the more memorable stops on our journey west. The first was in a grimy, third-class hotel in Waterloo, Iowa, where we arrived too late to be choosy. It was Connie's first experience with gray sheets, and she did not like them. It was a breathlessly hot night and we slept restlessly. Sometime before midnight I got up and looked out of the window to the street, faintly illuminated by a dingy light, and there, walking down the sidewalk, was an old drunk weaving his way toward a tavern from which faint sounds of revelry could be heard. About ten feet behind him, a large Toulouse goose followed, turning after him through the swinging doors. This was in 1936 and Konrad Lorenz's *King Solomon's Ring* did not appear until 1952. I had therefore never heard of imprinting, but this extraordinary goose behavior intrigued me greatly. It set me on a train of thought which, although I could not at that time follow to a conclusion, certainly conditioned me for my later interest in the phenomena of imprinting, bonding, and release mechanisms in both man and beast.

We paused briefly in Riverside staying with my grandmother and Uncle Bob. That evening at a large gathering of the Hellyer clan Connie, successfully hiding her trepidation, was welcomed, counseled, admired, and enthusiastically accepted into the family.

The next day it was old blue jeans, the dog, and westward ho! again. The familiar countryside unrolled before us—flatlands at first, hardwood groves, rich farmlands, then prairie, and gradually the scrub and sage of the high plateau country. We passed through Cheyenne, Laramie, the Wasatch Mountains, and down into Provo, Utah, then south through the increasingly hot and barren desert lands on into Las Vegas.

It had been four years since my previous ignominious ejection from that miserable city, but already its character had changed. A few modest hotels advertised their air-conditioned amenities, and it had lost the look of a raw construction town. The heat was intolerable so rather than driving on into the Mojave Desert we took a room in an air-conditioned hotel and napped gratefully for a few hours. Then at about 11:00 P.M. we checked out so as not to pay for a full day. We placed our suitcase in the main lobby with

Linda's leash tied to the handle, walked out into the heat of an open oven and on into the coolness of an all-night movie house where we spent three hours.

Returning through the quiet and nearly dark street we noticed that our hotel was blazing with light throughout the entire first floor and lobby. We had left the bags and dog under the main light switches and Linda would not allow anyone near them. We apologized, released Linda, and headed for the car. Unfortunately she scented another attraction, and instead of following us, she rounded the corner of the main lobby into the nearly deserted all-night bar—deserted, that is, except for an old sheepherder who was passing the late hours with a whiskey bottle while his crossbred border collie sat patiently, perhaps habitually, beside the bar stool.

Toenails skidding across the linoleum floor, and savage threats rumbling from her throat, Linda piled into the other dog. One of the most wide-ranging, noisiest, and least damaging dogfights that I have ever seen ensued—least damaging to the participants, I should say, for stools, chairs, ashtrays became involved, while the owner of the sheep dog spewed forth a stream of the most basic, colorful cursing, while remaining seated at the bar with his feet pulled up as high as possible to avoid the fracas. I was finally able to grab Linda and drag her to the car, and we were off, this time followed only by the expressions of relief and ill will instead of sheriff's cars. I have never been to Las Vegas since and would not risk it, as now the opportunities for misfortune, I understand, are virtually limitless.

We drove nonstop through the remainder of the dark hours and the incredible beauty of a desert dawn, arriving by midafternoon tired and with a feeling of unreality at Connie's family house in Santa Barbara. We stayed in the guesthouse as her family was away. My family was still back East. Dudley Carpenter was in Taxco, painting the pink sunlight and shadows in watercolors that live today in our house. He had gone there on the strength of my descriptions and fell under the magic just as I had, but he, with his artist's sensors and skilled brush, could capture it for others, while I must close my eyes to have it momentarily for myself alone.

It seemed a strange and lonely homecoming. It was probably a good thing that Santa Barbara revisited in this way was bereft of its usual ambience and associations, thus seeming more a part of the past. It was almost as though we had been granted permission to start north carrying with us little of the burden of nostalgia we had both anticipated. Instead we found ourselves excited to be on this last leg of our journey.

We followed the redwood highway and camped by a fine stream at the edge of a dark cathedral grove. For me, forest edges are much more appealing than the comparatively sparse understory and bare duff and penumbra of great conifer woods. I am fond of the chiaroscuro of a transition zone, of underbrush, tiny leaved plants, mosses penetrated by slender grasses, fungi and lichens, matted grasses where small creatures organize intricate mazes, and where sunshine evokes the latent scents. Deep woods awe. There is a little light or motion except in the upper story. Forest edges call to the small and poignant emotions while dark forests bludgeon the sensibilities. It was here that our long-considered decision to delay starting our family until the future became more clear lost its resolve—here at the forest's edge.

Northwest Beginnings

It was July 10, 1936, in the late afternoon that we reached Tacoma from the south, driving through the sprawl of South Tacoma Way, down the ravine and along Pacific Avenue—still a most unattractive approach, from which one cannot even imagine this city's beautiful setting on the bluffs above Commencement Bay. The Great Depression had not yet shown any signs of remission in the Puget Sound country. We could hardly imagine a more dreary scene than the boarded-up storefronts along the main streets. The narrow brass-trimmed stairways rising from the sidewalks on Pacific Avenue with arching round lights, carrying such names as The Dora, advertised that at least the oldest profession in the world flourished on the payrolls of Fort Lewis and the lumber camps. We settled into our temporary quarters in a tiny room at the Winthrop Hotel, initially dismayed at our prospects.

We thought of Santa Barbara, golden, clean, and familiar, and only two hard driving days away, but we did not mention it. Instead we changed and ate a mournful dinner in the gloomy, nearly empty hotel dining room. Then, because we could not face our room again so soon, climbed in the car and started down toward the tideflats. There we could see the glowing, fiery red waste-burner domes of at least seven mills. For the first time we inhaled the pervading and evocative scents of resins and woodsmoke, which were carried to us by an onshore breeze. We were fascinated and decided to explore that sprawling industrial complex in the fading light of the long northern twilight.

We crossed the Eleventh Street bridge, and turning right, found ourselves in front of a one-story office building with the name "Wheeler-Osgood" in large square letters above the wide entryway. Behind the building stood immense wooden sheds of plywood and door plants, while farther along the frontage, huge peeler logs were gathered, and rejected core blocks gleamed whitely in the gathering darkness. We circled the broad tideflats and saw the vast accumulation of logs stacked on flatcars, cold decked, rafted in the bay or in the ponds, and again savored the wonderful smells of resin, salt water-soaked bark, fresh sawed lumber, tidelands that saturated the air. Here was life, vitality, the heart and brain of Tacoma, and it was entirely new to us.

Reluctantly we recrossed the Eleventh Street bridge looking this time to the west, down Commencement Bay, and across to Brown's Point and Dash Point, where in the thirties, only pinpoints of light were scattered along the cutover and scarred bluffs. We spotted the dark outline of Vashon Island stretching halfway across the northern view. The deep indentation of Quartermaster Harbor was only faintly visible. The arms of the bay encircled the island on either hand—East Passage to the east and Dalco Passage to the west.

The city itself climbed the steep hills to the west. We followed the ascending contours past the great Norman chateau of Stadium High School, originally designed to become the Northern Pacific Hotel that would out-chateau Fontenac. We gained from this high point an even wider view of the complex waterways, the distant range of the Cascades, and very faintly, on the southeastern horizon, the great white ghost of Mount Rainier in the fading light. It was then that we first appreciated the unique beauty of this site. We continued to drive the treelined streets of the North End. There were mansions and small, modest homes, and everywhere lawns were green, and shrubbery, flowers, and ornamental trees lent a sense of permanence and caring.

Farther to the north the beautiful virgin forests of Point Defiance split the waters of the Sound like the bow of a ship, creating the swift current and tide rips of the Narrows. We went to bed that night, exhilarated, a little awed, but no longer looking behind us to the lotus lands of the South.

Early the next morning I presented myself at the Wheeler-Osgood office, where my arrival seemed to cause some embarrassment to the manager and curiosity in the girls who sat at their desks figuring invoices on their Monroe calculators. I was escorted around the office and introduced to everyone as Mr. Ripley's stepson. Only one or two of the staff remembered him, although as I got into the plant on other occasions, I found many old-timers who had tales to tell about him, and it was apparent that they thought of him with affection as a gentleman, somehow set apart from the other bosses, many of whom had come up through the ranks. It also became apparent that no one had the faintest idea what to do with me and that no real job awaited me. I was told that I would be paid seventy-five dollars a month, but that first I should go home, borrow a Monroe calculator, learn to use it, and come back in a week.

I spent mornings dutifully working through the instruction manual of my borrowed calculator. In the afternoons we continued our exploration of the city and its immediate environs. Although Tom Ripley had suggested various desirable North End residential areas, we decided from the first day that we wanted to get out farther into the country and find or build a home among the tall fir trees. Then, as now, we were not very good at accepting advice that ran contrary to our concepts, and so we ranged widely across the bay and up the Puyallup Valley. At last we discovered the lakes area, a few miles south of Tacoma.

Interlaaken, as the area was known in those days, would have been unrecognizable to the present Lakewood suburban resident. Most of the roads, with the exception of

First home in Interlaaken, Washington, 1938.

Gravelly Lake Drive, were narrow and unpaved. The three major lakes were not then completely built up, although the beautiful old country club was surrounded by homes, most of them suitable only for summer use, while the magnificent estates around Gravelly Lake were used year-round. Great rough Douglas fir trees, their heavy branches sweeping low to the ground as they do on the prairie, covered the area between the lakes except for openings where Garry oaks formed groves, and along the streams where big leaf maples, dogwood, vine maples, and other hardwoods flourished. Trails penetrated the entire area, and one could ride or walk to Chambers Creek or canter over the prairie where Lakewood Center now stands.

Connie and I had dreamed of this kind of habitat, and with the help of new-made friends we started to look for a house. There were not many possibilities, but after rejecting two small summer-style shacks, we found our future home. A large, brown-shingled frame house of two stories stood at the northeast corner of a ten-acre lot, hidden at the first floor level from the intersection of dirt roads by a collection of outbuildings in various stages of decrepitude. The house itself was soundly built, yet it was supported on blocks with no continuous foundation, which meant that water pipes were exposed beneath, and the wind played freely between the floor joists. The house had been empty for eight years. Windows were broken, plumbing torn up, wiring exposed, and the floor was a litter of broken bottles, fir cones, and the rejected Sterno cans, rags, and other detritus of hobo camps and boys' clubs. But under the litter were good floors, and the walls were smooth and ready for papering. The wide and welcoming front hall had a brick chimney to accommodate the stovepipe of a wood heater. The other rooms were also well proportioned and high ceilinged, and like many English country houses, there was charm and spaciousness, though little concession to modern conveniences. It had been built, in fact, by an Englishman, sometime about the turn of the century.

The lot, like the surrounding countryside, was shaded by huge prairie firs, but in front there was an open meadow choked with Scotch broom. Broom had been brought into the area as an ornamental shrub many years before and had spread throughout the coastal plain mostly by seeds adhering to the trousers of bicyclists who rode through its yellow clumps to admire the vigorous blooms.

So here was our house. For three thousand dollars we acquired ten acres, outbuildings and house included. The latter valued at five hundred dollars, and the Tacoma Savings and Loan Company agreed to either tear it down as derelict or to extend our mortgage to cover the cost of renovation. There was never any question in our minds of tearing down that house, and we decided what we would have to do to make it livable.

As I look back on the five years that followed, I wonder where we found the time and energy to do all the things we did. We had some help, of course. A diminutive Italian laborer, Tony Scapoli, helped clear the Scotch broom in the meadow. He also cut logs from our woods and helped us bring order to the dilapidated outbuildings. Mary and Fred Calvert, who lived in a neat little house near us, came into our lives at this time. Fred was a semiretired house painter and paperhanger. He was a frail looking man with the characteristic gray pallor of one exposed to lead paint for too many years. His wife was large, florid, and handsome. She did washing and household chores and taught Connie how to iron and fold my shirts. We met them when he was hired to paint and repaper our house, and from that time until his death the Calverts played a key role in our lives.

We moved into the house two months after signing the mortgage. The walls and ceilings were painted or papered, new wiring was installed throughout, plumbing was restored, and a series of old septic tanks, hooked up in tandem, were unearthed and quickly covered up again in the hope that their combined capacity would make up for their primitive design. With Fred Calvert's help, I soon built a little barn with two loose box stalls, a milking stall, and tack room with an overhead hayloft, and installed a Jersey cow and a horse I had acquired from a neighbor's daughter who was leaving for college. Another unemployed neighbor milked the cow on shares while I rode the horse through the woods, over the prairies, and onto the military reservation. There was also some urgent remodeling to be undertaken, for Connie was by now obviously pregnant, and in a mood for nest making. Accordingly, we converted the little room at the head of the stairs into a nursery with all the frills and appurtenances, and our first child, Constance Anne, was born April 22, 1937.

My work at Wheeler-Osgood was monotonous and obviously without future, but in those days, especially as I had been given a small riase, one did not abandon even a bad job without an alternative. There were some compensations, however, for I learned something about wood products, manufacturing, and established a love-hate relationship with plywood that lasts to this day.

Often I would go into the plywood plant and watch the

nearly flawless old growth logs unwinding against the lathe knife in a seemingly endless ribbon of slash-grain veneer. The sap squirted and ran from the logs and the aroma was wonderful. After clipping and sorting, the veneer went into steam-heated drying ovens, and when dry, the crossbanding veneer was covered with an even coat of glue at the spreader and the sheets laid one on the other in alternating plies of three, five, or seven layers, and thence into the presses. Waterproof plywood with phenolic glues was just coming into use. A small amount of the panel output was of this new type that required hot presses, but finally produced a product free of the delaminations that had for so long plagued the industry. Peeler blocks were the aristocrats of logdom in those days. Perhaps not more than 5 percent of the fine old growth fir logs would qualify. But now plywood manufacturers peel almost anything and call it good. But I must not dwell on a time that will never come again—when lumber was close-grained and beautiful—when a two-by-four was straight, nearly knotless and without checks.

An alley separated the Wheeler-Osgood office from that of the St. Paul and Tacoma Lumber Company. Founded in 1888, St. Paul's saw mill was one of the largest in the West and its timber holdings were extensive in Pierce County of which Tacoma was the industrial center. Chauncey Griggs, then secretary of the company, and I soon became close friends. He had been two years ahead of me at Yale, and we had attended rival California schools. Sometime during our early conversations I had told him that we were looking for a piece of land in some remote but easily accessible area on which to build a cabin. This idea seemed quite bizarre to most people to whom we mentioned it, for it was long before the vacation cabin fever had become fashionable. Most of our friends felt that our house in the woods ten miles outside Tacoma already represented an eccentric flight from civilization. But Chauncey was quite taken with the concept. We studied the maps of the St. Paul and Tacoma timber holdings, noting particularly creek frontages, lakes, and shores of rivers. We then set out to explore some of these sites, but only one of them seemed to fit our requirements, a small lake, designated "Horseshoe Lake" on the Pierce County maps.

About a week later on a sunny Saturday morning, Chauncey, Connie, and I set out on the old Mountain Highway, turning off at Johnson's Corner onto a dirt road, turned again just beyond Clear Lake and into a small, deeply rutted track, where we left the car. After walking about a quarter of a mile, we emerged into a clearing where neglected apple trees and an abandoned well marked the site of an early homestead. No evidence remained of house or cabin, but there was a meadow, which evidently had once been a cultivated field. Just beyond this clearing the track became a trail, dropping down into a shallow swale where Christmas treelike firs pushed their way above the thickets of underbrush and bracken. Just beyond a small, brown horseshoe-shaped lake seemed to slumber, so quiet was the surface of the water.

In the late 1930s the Horseshoe Lake property was just beginning to recover from the fire that swept through the area twelve years before.

The north shore of the lake, which would have formed the frog of a horse's hoof, was perhaps ten feet above the waterline, and six scarred fir trees up to eight inches in diameter stood on this promontory. They were fire marked but had survived the general conflagration that had swept through this entire area some twelve years earlier, leaving only a scattering of living old growth firs along the edge of the lake and on isolated hummocks in the swampier areas. Less fortunate trees stood at every hand—charred dead spars, already prey to lightning, the tunneling of carpenter ants, burrowing insects, the hammering and probing of woodpeckers, and the fungal and bacterial agents of decay. They rose above the ferns and the huge stumps of fir trees logged four years before the fire. Thick tangles of blackberry vines, willow, and small Oregon ash crowded the lakeshore above the muddy zone indicating a wide fluctuation with the seasons.

We could not guess how rapidly, or indeed how if ever, this discouraged landscape would be healed. Certainly the little lake with its stained waters in no way resembled the clear mountain gem we had half envisioned. No great pristine trees cast their reflections on its surface. But we were young, enthusiastic, and congenitally optimistic. This fifteen-acre lake was remote enough, yet close (only thirty miles) to Tacoma, and it was a bright and sunny place; it was also available—a hundred acres for four dollars and fifty cents an acre. In May 1937 the small brown lake on the plateau above Ohop Valley became Chauncey's and ours with the exchange of two checks for two hundred and fifty dollars each, made out to the St. Paul and Tacoma Lumber Company.

Immediately after purchasing Horseshoe Lake, we set about building a log shelter so that we might spend time there regardless of the weather. The entire basic structure was erected in one day with the help of a fine crew of friends. Connie, while keeping an eye on our infant Connie Anne, nourished us with an enormous Irish stew and lots of beer. We and our friends spent many carefree weekends exploring

The first house on Horseshoe Lake became a weekend gathering place for family and friends.

and even getting lost in the thick brushy swales and head-high bracken of the surrounding countryside. We swam in the warm, shallow water of the lake, made log rafts, and fished for the abundant smallmouth bass and the more elusive crappie. At dusk, we watched the catfish that mugged along the surface of the lake. They dove suddenly if one stamped a foot, causing the entire surface of the lake to explode with splashes in concentric rings. The summer evening skies were crisscrossed by soaring and swooping nighthawks, their underwings conspicuously barred with white. Their high-pitched cries were punctuated by the strange booming sound of air vibrating through the primary feathers of their wings as they dove. Male bullfrogs rumbled in tones whose depth of bass gave indication of their relative size and degree of libido, each stationed along the shore, holding to his own log or rock.

Yet not all our free time was spent in such isolated splendor. Because we were further along in life with our own home, a baby, and a more settled life-style than most of our friends, our house became a gathering place. A varied group met each Saturday night in our library. Someone jokingly suggested that we call ourselves the "Lakewood Bar Association," and the name stuck. Fueled by gallon jugs of beer purchased at "The Poor Man's Country Club" we talked endlessly, argued, for a time read the classics, and played music on our old Victrola.

Though personal and immediate problems or philosophical speculations were always our first topics of choice, the threat of war in Europe was increasingly on our minds. There were some in our group who did not believe that the German threat was any concern of ours. They claimed that it was merely another imperialist war and saw no difference between the Germany of Hitler and the Germany of the Kaiser. There were others whose sympathies were with the Allies, but sincerely believed in a strict neutrality for the United States. Still others, such as my brother George and myself, whose lives had been partly spent in Europe, were unqualified in our commitment to the Allied cause.

Discussions became heated and friendships strained as we all felt that our time of growing up, our time of innocence—when the sun shone on us and our lives were largely ours for the molding—might be coming to an end. There was tension in the air. Our peaceful existence was disrupted. Everything seemed tentative. It was as though a sinister black cloud was moving in from the East, casting a chilly darkness as it approached.

Early Skiing in the Cascades

My chance to change jobs came during the summer of 1937. Although there was a growing interest in skiing in the mountainous Puget Sound region, there were meager overnight accommodations except at Paradise Valley at Mount Rainier, where a fine lodge and inn had been built to accommodate the summer tourist trade and now to house winter sports enthusiasts. Still, there were no ski lifts in the state of Washington.

During the winter of 1936, Chauncey Griggs had spent many weekends at Paradise, and there he met Jim Parker, who had just come to Tacoma from Williamstown, Massachusetts, and was enjoying the deep snow of the Pacific Northwest for the first time. After repeated half-hour climbs to the top of Alta Vista above Paradise Valley, followed by minute-and-a-half downhill runs, Jim turned to Chauncey one day and said, "We ought to build a ski tow here. We'd make a fortune." Chauncey asked him if he knew how to make one and he said, sure, he had built one of the first on the East Coast at Williamstown the previous winter. And so, from that conversation, Ski Lifts, Incorporated, had its beginnings.

Jim Parker had been brought up in Europe, chiefly Switzerland, before his family moved to Williamstown. His mother, a widow, had written a popular book about their family life, so that everywhere he went and met new people,

Jimmy was identified as the cute, smart, attractive youngster of the story. This must have been both an asset and a liability. Whether it was this prefabricated notoriety or an inborn sociability and charm that smoothed his way through early life is difficult to judge. Be that as it may, when he arrived in Tacoma he was an instant social success and soon skiers and all who were interested in the promotion of winter sports knew him and were attracted to his plans for building ski tows on the snow slopes of the Cascades.

After he and Chauncey arranged for some financial backing, Ski Lifts, Incorporated, was officially founded in the fall of 1937. I had met Jimmy in the spring when he and Chauncey came to the house for a social visit. Their conversation was filled with enthusiastic plans and discussions of the complicated problems attendant upon the starting of their new venture. I was greatly intrigued, feeling that this was the sort of opportunity I had been looking for, to get in on the ground floor of an exciting project, to which my skill and energies might make a real contribution. Chauncey had insufficient time to work on the physical planning and construction side of the operation but had contacts and the financial know-how required to keep the project afloat. Jimmy had a good deal of experience from having built one ski lift already, but was not particularly mechanical or fond of monotonous tasks, and avoided, when possible, hard physical labor. He was, however, to a superlative degree, a born PR man, both through his gift of gab and his skiing prowess. These were the days of long skis, telemarks, and christies, although stem turns and snowplows were coming in, and Jimmy was poetry on skis. When he came down a slope, dipping and turning, he was magic to behold, and every man, and certainly every woman stopped, leaned on their poles, and watched. In the evening at the inn or lodge, he was always the center of attraction. Without Jimmy there to pave the way, there could not have been Ski Lifts, Incorporated.

But who, I wondered audibly, was going to spend time

at the two sites in the off-seasons, live in a tent or shack with a work crew while designing and constructing the buildings, setting poles, figuring out the tightening devices for the ropes, supervising the machining of the sheaves, devising safety gates, and making it all come together? And when this was done, who was to stand in the cold and collect the dimes? Or climb frozen poles with the weight of a wet rope on one shoulder replacing it in the pulleys when it jumped out in response to the bouncing and tugging of some high-spirited customer? And worst of all, who was to weave a long splice in a broken towrope while the lift stood idle and the dimes remained in pockets? Of course, none of us anticipated all these routine operating problems, but the construction requirements did seem to call for an additional partner, and I offered myself for the job, and became the third member of the company.

The principles of a rope tow are fairly simple, but in practice, when one is dealing with snow depths that fluctuate from a few inches to twelve or more feet, not counting drifts of twenty feet or more, and when the length of the tow is so great that the stretch and contraction of the rope may be more than thirty feet, ingenuity is called for, and I spent much time trying to solve these problems.

I gave up my job at Wheeler-Osgood with no regrets and embarked on this new career with enthusiasm. We built our first rope tow at Paradise Valley in Rainier National Park. The enginehouse was two stories high, to accommodate the tremendous snowpack, and was located just behind the inn, near a clump of trees, which afforded some shelter. The top pulley was placed on a pole at the summit of Alta Vista. Special conditions and restrictions were placed on our operation within the national park, since every evidence of human interference with the landscape had to be removed when the snow left the hillsides. Thus, each year we assembled the building in the late fall and disassembled it in the spring, masking the holes where the poles had been lifted from the ground with huckleberry bushes, and storing all the pre-

Ski Lifts Inc. comes to Paradise Valley, Mount Rainier.

ricated panels and machinery out of sight. This was a lot of extra work and required many days away from home at both ends of the season, as well as during the weekends of operation.

IN 1936 SKIING WAS in its infancy in the Pacific Northwest. Paradise Valley on Mount Rainier with its magnificent rough log and timber inn, its huge fireplaces, the less luxurious but less expensive lodge, and the small crude cabins offered by far the best facilities for the growing numbers of skiing enthusiasts. But it was hard work climbing to the top of Alta Vista or any of the other runs, and by evening there was little energy left for play.

Ski Lifts changed all that. It provided a rapid and sometimes exciting ride to the top of the slopes even though the weight of a wet rope and the spray of rope fibers that splattered clothing if your hands slipped were minor annoyances. No skier's subculture had been developed then. Skiers came to ski. They were largely self-taught. Their equipment—boots, bindings, even their skis—by modern standards would be grotesque. Tight pants and fancy clothes simply were not seen. There was no liquor in the park, but we provided our own jolly room parties. The ice for drinks came from huge icicles that hung from the roof in the two-foot space between the immense snowdrifts and the walls. Harry Papajohn provided food that still lives in the memory of generations of skiers. After dinner we danced schottisches in wool-socked feet, sang songs, and since the room keys were of a somewhat interchangeable pattern, flowed from room to room throughout the buildings.

There were no lobby skiers, no snow bunnies, no tellers of tall tales about world renowned ski resorts. When the snow was deep we parked our cars at Narada Falls and floundered two miles to the inn up the steep trail in deep snow, and skied back down in the half-light of evening through the thick timber of "Devil's Dip." We even instituted night skiing when the weather permitted and hung Coleman lanterns to the poles. We had difficulty collecting the dimes in the dark so many of these night runs were free.

For me and our lift employees the play was limited and it did impose an unwelcome form of bachelorhood. Yet for a time it was a way of life—a way that has now lost its naïveté to the sophistication that came with the new technology and popularization of the sport. Recently a group of prewar, pretow skiers formed an organization called Ancient Skiers Club. The membership is remarkably large and the get-togethers are said to recapture some of the jollity of those early years.

In the late spring when the snow was melting off the ski slopes but still deep behind the inn, I was allowed to stay in

Lines for the rope tow were long during the day, and the lodge and inn were packed at night as skiing boomed on Mount Rainier.

Paul Sceva's cabin (president of the Rainier National Park Company at that time) during the weeks of the dismantling process. The advent of spring on the mountain was exhilarating. The air tingled, and the alpine flowers seemed to spring up overnight at the edge of the melting snow. Where the long lines of skiers had stood waiting to catch the towrope, the glint of silver coins could be seen on the bare ground between the thrusting blades and stems. This was the small change that had dropped and sunk through the many feet of snow to the ground far below.

One spring, a mother bear with twin cubs hung around the kitchen door of the inn and the Sceva cabin. She was hungry after winter hibernation and from bearing and nursing cubs. In the mornings I could hear her scratching around walls and snuffing at the door. During the weeks when the inn was closed, the work crew ate in the kitchen, where Harry Papajohn dished out fabulous meals to the small workforce. I was the only member who was not living in the inn, and thus had a peculiar problem. In order to get to the kitchen door where the bear often stood vigil, I had to ski or walk down a steep slope of about a hundred yards. To hasten this short trek, I put on my skis in the cabin, opened the door a crack to see if the way was clear, and made a quick schuss to the safety of the kitchen. Unfortunately, the door opened inward, which made for difficult maneuvering on long skis. I developed an oblique technique that allowed me to slide through on an angle while pulling the door closed with a string. I never dallied on this early morning breakfast run. The return trip in the evening was much easier, for the bear was occupied at this time with the garbage.

After completing the Paradise lift we obtained permits to expand our operations to Snoqualmie Summit and Mount Baker, and I also built two "portable tows," which we thought could be used for special events at more remote sites. The ski tow at Snoqualmie Summit was of a similar design to that at Paradise Valley, but the third tow at Mount Baker was a far more challenging project. The most popular slope accessible to the lodge ran out onto a small lake making it necessary to place the enginehouse at the top instead of the bottom of the hill, while mounting the end pole on a raft in the middle of the lake and waiting for it to freeze solidly in place before installing pulleys and rope. Because this slope consisted of nearly solid columnar basalt, we had to dynamite the rock for the house foundations and the holes in which to set the poles.

One rainy, sleety evening, I was on my way with two helpers to relieve Jimmy Parker who had been on a fast inspection of the slope and the newly built enginehouse. We passed his truck, a secondhand highway patrol van, just as we started up the grade from Baker, and flagged him down. In a hurry to get back to warmth and civilization, he rolled down his window and shouted, "Everything looks fine at the tow site," then drove on. When we reached the fine stone cabin near the lodge that the Forest Service had loaned to us, it was nearly dark.

This cabin was built up against the hillside, had bunks, but more importantly, a large Monarch wood cookstove occupied one end and provided adequate heat for the building. We quickly brought in our sleeping bags, tools, and groceries, and lit a Coleman gas lamp and two kerosene lights. This made the place a bit more cheerful. We started a fire from kindling that had been thoughtfully piled to dry beside the stove, leaving the firebox door open to watch the flames catch and noticing the glaze of ice clinging to the stone wall of the cabin where it touched the hillside.

We then set about laying out sleeping bags and started to assemble cooking ingredients on the sturdy plank table that occupied the middle of the room. The fire was burning well and the door was now closed when we noticed a distinct smell of charred wood. Disturbed I walked over to the stove and opened the oven door. There, with wooden sides blackening and smoldering, was our nearly full case of dynamite sticks, caps, and wire that someone had either hidden there or stored there to keep dry. We looked at one another with horror. Then, not from bravery, but because it seemed the safest course, I grabbed our sleeping bags and bundled them into a thick cushion in front of the open oven door and then, oh so gently, I used the hook of a poker to draw the smoking box out across the oven door until it dropped onto the sleeping bags. I yelled for a bucket of water, but there was no one in sight to fetch it for me. I got the bucket myself, doused the case, contents, and sleeping bags, and then rushed out into the night. A few moments later, the three of us peered cautiously around the cabin door. Seeing everything was safe, we poured ourselves huge rations of bourbon and collapsed cursing the perpetrator of this insanity. When we did get home and angrily accused Jimmy, he remarked, not even sheepishly, that probably he should have left us a note so that we would have no trouble finding the dynamite when we needed it.

After the three ski lifts were completed, and the winter season over, the Mount Rainier enginehouse dismantled and machinery oiled and stored for another year, I found that I had time on my hands—an intolerable situation at that time in my life. Accordingly, I pursued a number of diverse interests during the summers, some bringing in extra pay, others providing satisfaction or new experiences.

In the first category was my association with Pat Erskine (his true name, Cyril Algernon A. Erskine, was never heard). Pat was an Irishman and proud of it. He usually wore a cloth hat with the brim turned down all around and a tweed jacket with leather elbow patches. His Irish terrier, Michael, went with him everywhere and ran behind him when he rode his Connemara mare. Pat came from the north of Ireland and had fought with the British in World War I when only seventeen, then emigrated to Canada where he learned the lumber business at its most basic level—cruising timber and scaling logs. He arrived in Tacoma in the late twenties and soon established himself as an independent log scaler.

Although some people resented and poked fun at his extreme anglophilia, which coexisted quite comfortably with his patriotic Americanism as a naturalized citizen, no one, buyer or seller, questioned the scale of a raft of logs Pat Erskine had certified. Knowing that I was looking for odd jobs during the summers, he took me on to tally for him. Whenever he had a scaling assignment, whether on the Sound or even as far away as Vancouver Island, I would go along and help. Sometimes we stayed in logging camps, eating the enormous silent meals, for social chatter is never welcomed. More often, the jobs were close at hand, and we went home at night. I walked each floating log with him, back and forth over the big rafts, as he strode with measured tread, scaling rule rising and falling until he reached the end of a log, then hooking the underside, measured the diameter of butt and top. He could tell the species of tree without glancing down by the way his caulked boots bit into the bark, and he could detect the hidden flaw or jagged limb below the waterline by the way the log wallowed in the water. Deftly rolling it with his feet, he could even determine the extent of the defect. Occasionally he followed these logs through the mill to see how accurately he had scaled and graded them. For it was on this accuracy that both buyer and seller had substantial sums to lose or gain. I would often come home exhausted after a day on the logs, but Pat, nearly fifteen years my senior, seemed tireless as his right forearm, nearly twice the diameter of his left, rose and fell rhythmically with the heavy rule.

Through Pat Erskine I met a young veterinarian, Louis Todd, and we became good friends, sharing an intense interest in animals and their care while studying their natures and behavior. I spent as many hours as possible in his clinic, at first assisting him in surgery and later performing the simpler operations and procedures. I particularly enjoyed the cow and horse calls where one had to work in barns or fields, learning a great many useful things about the common ills of domestic stock. But like all the other activities I undertook, this did not seem to engage my full interest or lead to a firm commitment.

It was a period of multiple and seemingly random activity, although in retrospect, from Ski Lifts I learned much about construction, developed my know-how, and gained an even greater appreciation of the high country. From my veterinary experience I gained knowledge of both the problems of handling ungulates and the peculiarities of their anatomy and behavior—all immensely important assets in later years.

A New Calling

WE HAD MANY VISITORS during those first years in our Lakewood house. Our families came—three generations of them—and they all seemed to approve, although it was sometimes hard to know what they really thought. Mr. and Mrs. Cate came several times. It was a fine thing to spend evenings informally and warmly with a headmaster one had admired but viewed with awe for he was known as the "King" to all his students.

There were others, both contemporaries from the school or college, and also a few of an older generation that had been friends around the tea table in Santa Barbara. The first of these was Dudley Carpenter who visited us soon after we moved in. He had returned from his painting trip to Mexico and brought us three watercolors of Taxco. They sparkled with the brilliant hot light reflected from white and pastel surfaces, which contrasted with the shadowed thick-walled arches and doorways suggesting coolness and siesta. During the days when I was at work, he painted our October woods and lakeshores, enjoying the serenity of muted tones, subtle gradation of greens, and the spiky outlines of conifers that were such a contrast to the brilliant Mexican colors and blunt forms.

He and Connie talked at great length about our plans and my job, with its obvious lack of future. He felt strongly that my capabilities could be directed into a career in medicine. At that time, when our life together in Tacoma was just beginning, the idea seemed too remote to be seriously considered. But three years later, when my involvement in Ski Lifts was decreasing and I was turning to other interests, especially veterinary work, we recalled Dudley's words.

We both remember well that November evening in 1939 when Connie quite suddenly began to question the direction our life together was taking. I listened with a mixture of rising excitement and vexation. In one way her words seemed a criticism, but in another, offered prospects of a new career I had never consciously considered, yet which seemed to be emerging from some subliminal depth, terrifying in its improbability, yet altogether right.

I realized what an important and respected role doctors had played in my life. I thought of Dr. Martin in Wales, making his calls in his battered old car—often to farms where he would have to finish his journey in a buggy or wagon with his huge old medical bag balanced on his lap—and the warm respect and affection that greeted him everywhere. I remembered Dr. Lancaster, the father of one of my Sunnydown school friends, who took care of me when I had a "septic" leg and later treated my rheumatic fever. An amateur ornithologist, he always took time to talk about my egg collection and other natural things. By contrast to these rural kindly men, I thought of Dr. Patterson, imposing, courtly in his Harley Street morning coat and striped trousers—the very image of success and wisdom as he listened to my chest and moved my painful, rheumatic joints with gentle, sure hands. In Santa Barbara there was Dr. Manning, who carried me in his arms to his car and the hospital when I had been found half conscious in the tall grass of my special place on the mesa with acute appendicitis. He, too, made a friend of me.

But most important of all had been my relationship with Dr. Charles Aylen of Puyallup, who delivered all three of our children by caesarean section, complicated by a variety of medical problems for their mother. During those long hospitalizations I naturally spent many hours in the tiny hospital where Dr. Aylen practiced. He was a shy, but warm and tender man who understood my worry and distress. He talked to me about his cases and gave me permission to spend time in the hospital playing with my babies, which was considered bad practice in those days. He was also interested in my future. (Later, when I started my premed, he called me, delighted, and for many years kept track of my career, assuming—quite rightly—that somehow he had been a sponsor.)

Connie and I talked long into that night, mostly about

Connie with Dorothy, and David with Connie Anne, Christmas, 1938.

the obstacles, as well as the reactions of families and friends to such a radical change of course, which would sacrifice so much of what we had built in exchange for a future of great uncertainty. We realized that I had studied none of the required subjects in premedicine except zoology and biology—no chemistry or physics—I did not even know what else I lacked. We tried to judge how such a change would affect our two children, Connie Ann, now two and one-half, Dorothy, only thirteen months younger, and the third baby just under way. We talked of leaving the home we loved and had worked hard to make warm and comfortable. We talked about our friends, but most of all we found ourselves reluctant to leave the lake, a growing source of pleasure and the focus for many future plans. But despite all these concerns, we both had a sense of exhilaration, a knowledge that now that this option was before us, it could not be ignored or set aside without full investigation.

We agreed not to discuss the matter with anyone until we had found out more about its feasibility. The following morning I drove to Seattle and presented myself at the Office of Admissions at the University of Washington, asking to speak to someone who could advise me on requirements for a premed major. I soon found myself seated opposite a professor of physics who was an adviser to students preparing for a career in medicine. We discussed my past academic record, my present status, and he explored sympathetically my still rather amorphous motivations. Apparently satisfied, he asked if I was really willing to set aside all my other interests and activities and work unstintingly to reach my goal. I choked a little but declared I was. As a result of this interview I went home that evening with a tentative schedule starting the following January that included three chemistry courses to be taken simultaneously, as well as physics, psychology, embryology, and several other courses. All this would be crowded into an eighteen-month period leaving summers free to work to replenish the exchequer.

Two days later Connie and I drove to Portland to discuss our plans with the Admissions Department at the University of Oregon Medical School. (Washington had no medical school at that time.) And there again, our reception was cordial and sympathetic. The registrar and two staff physicians talked with us for four hours, taking time out for lunch together in the hospital cafeteria. They examined my schedule with raised eyebrows and wondered out loud if it was possible to condense so much in so little time. But before we left, I was virtually assured that if I followed through and did well with my grades, I would be admitted to the freshman class of the Medical School beginning in September of 1941.

I had not expected such interest or depth of questioning, nor the invitation for Connie to join in the interview. I am sure that her presence was crucial. Ordinarily such interviews are not granted to prospective students even before they have embarked on a premedicine course, and certainly the conditional assurances that gave us the go-ahead were remarkable—I did not know how remarkable at the time.

We drove back home that night, our heads and hearts awhirl. The next day I returned to the University of Washington and found I could enroll in some of the required courses without waiting until January. I rented a small room near the campus above a grocery store, which boasted a two-unit gas burner and an apple box on the window ledge for a food cooler. The toilet water tank served to preserve butter and eggs. Late that afternoon, when I arrived home, we sat down somewhat breathlessly to evaluate what we had wrought.

Our position would be financially precarious after starting medical school, but during the premed phase I would continue working part-time on weekends with Ski Lifts. We decided, however, that we should sell our interest in the company to the other partners to help carry us through the two years before medical school and start accumulating a backlog. We did not realize much from this sale, but when added to the three hundred and forty dollars monthly income Connie and I now shared, we felt that we would get by financially.

When we had finally and somewhat defensively announced our decision to our friends, we were surprised and delighted by their enthusiasm and support. Our doctor friends were particularly encouraging, and best of all, my brother George was immediately enthusiastic, expressing his certainty that this was the career for which I was suited both by temperament and scientific bent. With Connie's staunch support and conviction that I would not fail, and with the backing of all our friends, I felt truly trapped during that first year of premed.

My first months at the University of Washington were difficult in a number of ways. I was older and married with children, which set me apart from my fellow students. I was treated accordingly but with some skepticism by my young instructors. I had also lost some of my study habits and found it difficult to absorb the enormous volume of material that had to be memorized, especially since I was simultaneously taking the several courses in chemistry that should properly have been presented in sequence.

I must emphasize how pervasive and terrifying my fear of failure was. I had deliberately turned my back on a fairly secure and most pleasant life-style that promised a healthy environment for our growing family. All this Connie and I had decided to exchange for an immediate future of uncertainties and, to me, possible failure.

Most nights of the week I stayed in my room above the store, studying late and working in the labs when permitted after-hours. I bought what perishable food I needed for a day ahead and, at weekly intervals, cooked a roast in my Dutch oven on the gas burner. This made meal preparation quick and easy between the weekends. On Friday nights I drove home and when the weather was clement we went to the lake. If I could manage the time I continued to work for Ski Lifts during the busy winter season.

When I received my final grades in June I could hardly believe them, for they were far better than my grasp of the subject matter warranted, reflecting, rather, the result of desperate rote memory. But that did not matter. My confidence was growing, and the recurrent nightmares of being refused admission to medical school largely disapppeared. That report card with all those lovely symmetrical pointed little mountains I had apparently climbed was there to gaze upon in wonder and became the reality.

When the time came for me to apply to medical schools, I returned to the University of Oregon with my Yale records, together with my new premed transcripts. This time I talked with the Director of Admissions. He asked me if I had made application to any other medical school, and I told him that I had applied to Yale because it was my undergraduate school, where I had made a fairly good record, and to the University of Chicago because I had heard so much about its liberal approach to education. He was most understanding, pulled out a file, which had apparently been started when I had paid my first visit a year before, and reiterated that I was sure to be admitted to Oregon if my grades remained at their present level. Then he said quite suddenly, "I went to the University of Chicago. It is a wonderful school—perhaps you should consider it quite seriously if they will accept you."

Up until that time, Connie and I had thought of the applications to Yale and Chicago as insurance should Oregon fail us in some way. After this interview we began to analyze our motives for wanting to stay as close to home as possible, for renting our house instead of selling, and for maintaining the old and safe associations. Were we not embarking on an entirely new venture, while in fact sitting in the gondola of a firmly tethered balloon? We began to wonder whether our desire to spend as much time at the lake as possible, our need to share our anxieties and triumphs with our friends, and in fact all the lures and distractions of home, might jeopardize the whole project. We both concluded that to cut our ties for the duration of our training, while keeping as our goal an eventual return to the Northwest, was the wisest, even if the loneliest, course.

And so we waited now with open minds for the replies to my applications. The first was from Yale. It was a neatly typed, admirably brief letter. I did not keep it. It stated that the experience of the medical faculty with older students who had decided on a medical career after trying other occupations had been unsatisfactory in most instances, and they rejected my application. Connie and I read and reread this letter with a sinking feeling, for I had been almost certain that Yale, if anywhere, would show some interest.

Fortunately, it was only a few days later that the letter came from the University of Chicago. Connie opened it while I was in Seattle. It was quite a long and personal letter, commenting on the fine supporting recommendations that my biologist friend, Richard Bond, and author-naturalist Donald Culross Peattie, had written on my behalf, and suggesting that my varied background and family status might indeed be assets rather than liabilities. When Connie came to the words "accepted in the class of September 1941," she could hardly read the concluding formalities through her tears. Somehow she reached me by phone in a lab at the university. I came right home. We celebrated with anyone who would listen, and from that moment our decision was firmly made. We were going to Chicago.

Clearing the Decks

YES, WE WERE GOING to Chicago. We realized that a far more radical change in our plans was necessary than if we had gone to Portland within easy reach of home. We decided to sell our house to provide funds for the coming years of medical school—tuition, books, a microscope, and other supplies. Although the lake property provided no income, it cost very little to maintain. We now owned the entire one hundred acres ourselves as Chauncey had turned his share over to us for two hundred and fifty dollars, the sum he had originally invested. For eight dollars an acre, we also bought an additional eighty acres through which our access road passed, which included the old homestead, a well, and an open meadow. So we focused our resources, future planning, and dreams on the lake property and decided to invest in a permanent cabin that would serve a dual need—a place to live during the coming summers when we assumed medical school would not be in session, and a secure future base for that part of our lives that demanded woods and water, wildlife and wilderness when we were no longer entombed on the South Side of Chicago.

Early in the spring of 1941, shortly after our fateful decision, we started building the new cabin on the site of the old log structure. We built the walls of hollow concrete blocks, twelve by four by six inches, made by hand eight at a time on a vibrating table—the invention of a Mr. Michaud,

A permanent cabin on Horseshoe Lake became a sanctuary when family life demanded woods and water, wildlife and wilderness.

who had previously built the brick fireplace and chimney for the library of our house in the woods. He had allowed the bricks to cure for several weeks before laying them up with a rough but sturdy mortar line, peppered with cigar ash and seasoned with pungent comments on the times and frailties of man. The floor and six-foot deck that extended the length of the front of the cabin was a concrete slab. Across the west corner of the twenty-seven-by-seventeen-foot building, a large raised fireplace, made of the same concrete block capped with a sloping rounded masonry surface above the mantle, merged with the walls and exposed chimney. The southwest corner of the cabin was made bright by wide corner windows. French doors and a heavy Dutch door completed the front facing the lake.

The interior of the cabin was finished by partitioning

off a tiny future bathroom, which was tucked behind a curtained double bed raised four feet above the floor and recessed between the masonry of the back wall and a wooded partition containing shelves and cupboards. A massive black iron Monarch wood stove decorated with German silver, flanked by a sink set in a heavy plank counter, fitted neatly into the east end of the cabin. A small window above the sink helped light this area, and overhead a twelve-by-sixteen-foot loft was reached by a ladder. There our little girls had their beds, played, and were able to spy on kitchen activities through knotholes in the planking of the floor before we covered it with an old carpet. I had helped only sporadically with the construction because I was still busy on weekends dismantling the Mount Rainier ski lift and commuting during the week to classes in Seattle. But as soon as the masonry and basic carpentry were completed we moved in and turned over our house in the woods to its new owners.

My premed work was satisfactorily completed in June. We applied for and were assigned housing near the University of Chicago campus. For the first time since our marriage we had time to ourselves as a family, and for me, no feelings of guilt for tasks undone or obligations unfulfilled. The knowledge that we had successfully climbed the first rungs of the academic ladder while managing to arrange our affairs so that we would leave Tacoma in good order and with some financial security gave us a sense of real accomplishment. At the same time we had made sound plans for maintaining, through the cabin and lake property, a firm foothold in this beautiful Puget Sound country to which we were now committed to return when my training was completed.

We had arranged for the Calverts, who were now retired and most anxious to get out into the country, to move into our cabin as soon as we had left for Chicago. The long-term plan was that they should live there, keeping Linda while we were gone. At his own pace, Fred would build a barn out of the down cedar on the property, then later start on a small caretaker's house for themselves up by the homestead at the entrance to the property. Later he and I would fence the land and get some dairy cows from which he and Mary might gain some income by selling milk and cream. In the meantime, we were to furnish what materials he needed and a bare subsistence wage. It was a wonderful boon to both families, for frail as Fred appeared, he worked methodically and skillfully all day, and soon he had started on the caretaker's cottage. We were truly interdependent friends during those years, who honored each other's differences and skills while recognizing each other's dissimilar backgrounds. Both Fred and Mary said before Fred died in the caretaker's house in 1953 that the years at Horseshoe Lake were the happiest of their lives. For us, their care, affection, and their almost single-handed building of the caretaker's house, the barn, and the peripheral fence, were the indispensable basis for our future ranch operation.

What a fine summer that was. Our friends came often to see us, sometimes camping overnight. Brother George and Babbie, with their children, Marion and Harold, were frequent visitors. When guests stayed for supper, it became traditional to take a walk, starting just as the sky darkened and returning after the trail could no longer be seen but only felt. No flashlights were allowed, and I usually led the way, showing great outward confidence that I knew where I was, but in truth, often confused for a time until a familiar crossing or silhouette against the night sky gave direction.

Once a week we took our Chevrolet Suburban carryall to Eatonville to shop for groceries at the lumber mill's company store, where they stocked many Oriental specialties. The mill was almost exclusively manned by Japanese labor. They lived in their own housing, had their Japanese baths and recreation near the mill, and seemed to live a happy and congenial life in close association with their logger neighbors of European descent. All was serenity in the little town of Eatonville, population about eight hundred, when we left for Chicago. Yet only a few months later Pearl Harbor was bombed, and overnight Japanese schoolmates and friends were the enemy. They were moved out with only what they could carry to relocation camps while their cars and property were bought at forced sales for a song. No Japanese have returned to Eatonville.

During the summer of 1941, however, we did not worry about a Pacific war. The battle of Britain was at its height. Young RAF pilots were showing up at McChord Field to learn to fly Boeings before ferrying them back to join the desperate battle. They were fine young men, really boys, mostly from the well-educated "upper classes" who knew that their chances of survival were small.

We came to know a few of these fliers well, and one day they came to a party at the lake with their dates, who were young teachers from the Annie Wright Seminary in Tacoma. We had a keg of beer, which sat on a stump in front of the cabin, and most of the "Bar Association" was present. The fliers really let their hair down. No diplomacy was required. They played on the log rafts and fell in the water fully clothed. We sang, told stories, and were free, but running just beneath the surface of the revelry was the sure knowledge that this was an isolated moment in time and that soon these boys would be in deadly combat in the air over Britain. Indeed, Pat Erskine, who kept track of most of the ones we knew, said that by the end of the war only two of the ten present that day had survived.

That summer I also had a chance to acquaint myself further with the natural riches of the property. I often saw black-tailed deer, but more often heard them move out through the brush ahead of me. I learned where they traveled, fed, and bedded down. We occasionally saw raccoon, mink, muskrat, and weasel. We found a pack rat nest—a pile of sticks decorated with stones, dogwood buttons, and various bric-a-brac—built around the trunks of a small clump of dogwoods quite close to the cabin. It had the characteristic sharp, urinous smell that is familiar to anyone unfortunate enough to have had his cabin invaded by a member of this species. Bear were, and still are, quite common in this area. On one

During the summer of 1941 British fliers spent a carefree day on the lake before going off to do battle over Britain.

occasion I came face to face with one at the site of an old homestead. It was a large cinnamon bear that was busily eating wild blackberries and riding down the branches of the old pear trees to get at the small hard fruit that burdened them. We saw each other at about the same time, and I went east at a very good clip, while the bear went west.

To me, the coyotes were the most fascinating of our neighbors. Until we had come so recently to the lake, it had been an isolated area, rich with small prey, grouse, snowshoe hares, mountain beaver, and smaller rodents, as well as deer. The number of coyote droppings and marking places where clumps of grass grew thicker and scratch marks were evident told the story of their crisscrossing travels and favorite hunting grounds. Of course, tracks in the mud by the lakeshore revealed the presence of other rarer visitors, such as bobcat and an occasional cougar.

At night, particularly when the sky was bright from stars or moon, coyote songs seemed to come from all directions. They yipped, laughed, and crowed like roosters from some distant hill. They also moaned and howled and when singing a duet, sounded like a load of excited children piling out of a school bus after a long and tiresome ride.

This special summer sped by all too quickly. The night before our departure for Chicago there was a gibbous moon reflecting shimmering trails of silver where fish mugged along the quiet surface of the lake. The coyote chorus that night seemed to me particularly wild and sad and joyous, attune with our emotional good-byes the following morning in September 1941. George took us to the train, other friends waved us off, and more than one throat struggled with aching lumps that could not be swallowed and tears that stung but would not flow.

For us the world had now become two places and two ways of life—the lake with its wild surroundings, its birds and beasts, and Chicago with its offering of knowledge and an opportunity to become a physician. The first life had to be sacrificed for the time, but we were certain that if we could accomplish the second, a single way of life that was a blend of both would emerge—and so it did.

Medical School

THE UNIVERSITY CAMPUS was beautiful, and the buildings just the right blend of function and ivy-colored college Gothic. An ambience pervaded the shaded courtyards, walks, and halls that spoke of scholarship, standards of excellence, and intellectual ferment. I liked it, feeling that what I needed to discover and learn was here for the taking if I had the will and the ability.

I was at once deeply caught up in my studies. I was fascinated with anatomy and physiology, and the smell of formaldehyde was never completely absent from my clothes, skin, and hair during that first year. Chemistry, however, was another matter. No matter how hard I worked and memorized, I nearly failed this subject. We had scarcely fallen into the routines of our new life when, with the suddenness of a tornado and the shock of an earthquake, came Pearl Harbor—and the United States was at war. Within a few days we were called to a convocation and informed that so long as we did well as medical students, we were ordered to remain frozen in that role. There would be no summer holidays, and we were expected to finish our four-year curriculum in three years. The dean pointed out that there must be no dropping out or volunteering for the military. Gone was the idea of summers at the lake, with the opportunity of earning money to help with the coming year.

Only a month before our entry into the war, Connie

and I had received a letter from the St. Paul and Tacoma Lumber Company suggesting that, since we already owned a hundred acres in the heart of section thirty-five, they would be willing to sell us the balance of their holdings above the Ohop Valley consisting of some three hundred and fifty acres (plus or minus) for the same monthly payment, extending our contract over a twenty-year period. This offer was irresistible. We signed and returned immediately the enclosed agreement, thus deepening our involvement with the lake property and our determination to return to the Pacific Northwest.

The plans we had made with the Calverts and certain other tentative business arrangements would now have to be modified and firmed up for the long pull. Accordingly, we decided that I should take a quick trip home during the two-week Christmas break to settle our affairs. Fred Calvert met me at the railroad station in the carryall we had left for him and drove at once to the cabin. The weather was unusually cold but clear. It was wonderful to be back. I slept in the loft as the Calverts used the big built-in bed, and while Mary kept the house warm and provided delicious meals, Fred and I started building fences, sawing and splitting posts from the many cedar snags that still stood completely dry and sound, more resistant than the firs to decay.

Before leaving for Chicago, we had bought an enormous, gentle logging horse with the appropriate harness, traces, and chains. His job was to yard logs out of the woods, pull a stoneboat with posts and wire for fencing, raise barn timbers, and provide for all our nonmechanized horsepower needs. His name was Bill, and he must have been mostly Clydesdale to judge by his fine, bright bay color, white socks, and feathered fetlocks. When he was not working, he roamed freely near the cabin, coming when called for a bait of grain. I have known many horses, but none more useful or cooperative than old Bill. Long after we were fenced, roads built, and farm machinery acquired, Bill had the full run of the place to the end of his days.

Bill, the enormous logging horse, was Fred Calvert's right-hand man when it came to hauling fence posts.

It was a wonderful two weeks, working long days in the cold, clear air, setting posts and stringing wire, then returning tired to the cabin, warm with fires in the cookstove and fireplace, the smell of cooking filling the room. I could scarcely bear the thought of returning to the Chicago winter, and felt guilty at enjoying our beloved lake without Connie and the children, although they were with family spending the Christmas break with my grandmother and Uncle Bob in Riverside. Before I returned, Fred and I bought two milk cows that had just freshened so that there would be milk, butter, and cottage cheese to help them through the lean times ahead.

That first Chicago winter was really rough for all of us. The children contracted one infection after another due, partially we believed, to the crowded conditions of city life, which must have overwhelmed their largely underdeveloped resistance to infection. There was a strong psychosomatic component as well, for they really pined for the more spacious life they had known.

Connie Anne felt it most of all, because she was the oldest and certainly the most restless and adventurous of the three. We finally discovered that by allowing her an entire wall to draw on, she could work off some of her frustrations, and perhaps at the same time maintain her health. Later she recalled it as her "Wailing Wall." "Doro" also had trouble coping and began to look thinner, paler, and more listless. One of the staff pediatricians, who had been following them in the hospital clinic and was always willing to call at our apartment, finally said that he felt that something would have to be done. He asked if there was any way that Connie could take the children back to the Northwest where they had been so healthy.

Of course, there was only one place for them to go—the lake cabin. So again we were to be separated, this time from the beginning of March through July. When Connie arrived at the lake with our wan little girls, Mary Calvert clucked over them while disapproving of the medications they had been placed on prophylactically. Somehow Connie and the children packed themselves into the loft, and they all managed as one extended family.

At once Fred started to build a little shed up at the homestead. This was to be the milkhouse and later a small oil and storage shelter. It was a tiny building—ten by twelve feet—with a little covered front porch. When it was finished a month later, the Calverts managed to crowd themselves into it together with their double bed, small table, and woodstove, so that by April Connie and the children had the cabin to themselves. I had built a sizable storage woodshed against the blank back wall of the cabin. This was well stocked with wood for the fireplace and blocks to be split for stovewood. Linda loyally stayed at the cabin, although she paid frequent visits to the Calverts in their temporary home up the road.

Connie wrote to me almost every day, describing the passing seasons at the lake, the almost miraculous response of the children to the familiar outdoor life, the coming and going of the wildlife at her doorstep, and reporting on the progress of fencing, the beginning of the barn, and finally the start of the Calverts' caretaker house. I almost felt that I was a participant, so detailed and intimate were those letters.

Meanwhile I tried to adjust to bachelor life. Actually, I was so busy that although lonely, the time did pass rapidly. I ate my lunches in the hospital cafeteria but cooked breakfasts and dinners, thereby saving money and enjoying more palatable meals. In the evenings I studied, using flashcards as

Caretakers, Fred and Mary Calvert watched over the growing Hellyers and the Horseshoe Lake property for many years.

a learning aid, for in those days we memorized every bone, every muscle origin, insertion, and innervation.

Involved as I was in medicine, I did not lose my interest in natural history. In fact, the more I learned about anatomy and physiology, the more the two fields became integrated. When possible, I haunted the magnificent Field Museum—surely one of the world's finest museums of natural history. When my work load permitted, I visited my grandmother and Uncle Bob on weekends, filled up on good food, and often borrowed a car to visit the Brookfield Zoo across the river, so close one could hear lions roaring. I became friends with the curator of mammals and was allowed access to the inner working areas of the zoo.

The war was on all our minds. I carried a sense of guilt that medical school exempted me from active participation. We had been told that if we kept our grades up, we would soon be given the opportunity to volunteer for enrollment in either an army or navy program. We could commit ourselves to various periods of active duty, but this was deferred until graduation. This helped solve the guilt problem. When the time came I selected the navy and was commissioned an Ensign HV (P)USNR on March 20, 1942.

Anatomy, Physiology, and the South Side of Chicago

Connie and the children returned to Chicago in the middle of July, and what a great reunion it was, although we realized almost at once that our quarters had become impossibly crowded, especially in the humid and stifling heat of a Chicago summer. I had become accustomed to having the apartment to myself and to studying in seclusion, but now with the children larger, more active, and used to space and uninhibited self-expression, our living quarters seemed to bulge and reverberate. Connie found another apartment on Seventy-second Street, some distance from the campus. Nothing stood between the building and Lake Michigan except a stretch of grass and a paved pathway. The air blew fresh off the water, and there was room for the children to play out in front and away from the traffic.

None of the apartment leases in that area of Chicago permitted dogs, cats, or birds, but they did not specifically exclude exotic pets. We decided to test the rules by acquiring a de-scented skunk. It was not a very satisfactory pet, but the children liked it as it neither bit nor scratched, but simply looked resigned and miserable when handled. During most of the day it lived under a radiator, but prowled the apartment at night and proved itself a useful exterminator as we were little troubled by mice or cockroaches. Even if not a responsive companion, it provided a faint scent of wildlife in our urban environment. When we moved again, we released it in the forest preserve, since it had proved itself a superior hunter of insects and small rodents. And so the summer and fall passed.

We had one idyllic weekend away from the city when my Aunt Edith offered us her summer home on the Rock River, together with the loan of a car to get us there. Several members of my family worried about us living and working so continuously in the city without any break, but had by this time accepted that I was serious about medicine and that we were not about to ask for any help. This offer, however, was too good to refuse.

I had never seen Aunt Edith's "cabin," a euphemism, for it was the most magnificent, two-storied log lodge that can be imagined. It stood on a rolling knoll overlooking the river and was surrounded by one of those marvelous prairie groves of various hardwoods. As it was fall, the trees were an autumn palette of umbers, sienna, chrome yellow, and shades of red. The refrigerator was crammed with food. The bar was stocked. The hot water was turned on. There were new and frivolous magazines and books on the coffee table in front of the huge stone fireplace, where a fire was laid with hardwood logs, and an alcove at the side held a reserve stock of wood. Even the beds were turned down.

We explored first the house and then the grove and riverbank. After drinks of aged bourbon, we ate nearly forgotten delicacies for dinner, followed by real coffee and exotic liqueurs. The beds swallowed us in soft down, and when we woke it was to find that it was not a dream.

Close family friends had a lodge very similar to Aunt Edith's, just up the river and around the bend from where we were staying. I knew that they had a farm where they raised fine Arab horses. I walked over to the barns that morning and introduced myself to the farm manager, who had been told that we were going to be neighbors for the weekend. We talked horses while he gave me the grand tour of the stables. When we finally reached the loose box stall and paddock where the stallion was kept, he casually asked me if I would like to ride him, apparently aware of my experience with horses. At first I demurred, but he said the horse needed exercise and that I could try him first in the paddock, then, if we got along well enough, I was free to take him out on the trails that were laid out in the surrounding woods.

I had never ridden such a horse. He was chalk white with an indescribably proud crested neck and mane, small, delicate ears, and a dished Arab face that set off flaring nostrils. As I sat on him, I could feel between my legs the quivering of hard, eager muscles, and he responded to a touch of heel like a taut string. He was a gentleman, well schooled, and I took him through his paces in the paddock, slowly at first, then the manager opened the gate and we were off for one of the best rides I have ever taken—especially magical after the long hiatus.

It was difficult for all of us to go back to town, but we were refreshed and everlastingly grateful for the gift of those three days in a natural setting at just the time when we as a family needed most to be reminded of gracious living, of leisure, and of its uses.

In May 1943 we students began our clinical clerkships. For the first time, after two years of preparation and anticipation, of memorization and theory, we put on a white lab coat, bought an otoscope and ophthalmoscope, and were tempted to acquire the fanciest stethoscope available. Then with what confidence and bravado we could command, we faced our first patients. There was a pretense by both student and teacher, and an acceptance by the less sophisticated patients that we were indeed doctors—very young looking, perhaps, but called "Doctor" by professor and hospital paging system alike, and we were expected to act the role.

The art of history taking, which we were supposed to have learned in class, suddenly seemed immensely complicated. We felt as though the patient were deliberately attempting to frustrate us by his apparent evasions and obfuscations. When we finally began the physical examination, we realized our awkwardness and were conscious of the tremor of our hands. In our eagerness, we searched for abnormalities in every breath and heartbeat, often finding them. Abdomens, when palpated, were full of questionable areas of resistance, or were they masses? Moles—perhaps melanomas? The anxieties we may have unwittingly transmitted to our patients became in our eyes signs of their neuroses, and we recorded all this on the charts.

Finished at last, we excused ourselves to present our findings to our mentor over a quick coffee or Coke. As the moisture dried on our palms, we tried to organize and present our findings in the approved manner. We may not have been rebuked for forgetting to ask the patient's age, and we might even have been gently chided for having omitted the past history or even the present complaint—but only this first time. An apparently casual checking of our "positive findings" was performed with consideration for our feelings, again only the first few times, and that is how we learned.

We were taught by giants then, and because of the small size of our classes and the full-time commitment of our professors to teaching, we knew them well as revered friends. The names have resonance to those who absorbed their knowledge and sensed their humanity all those years ago, and even though many are gone, their contributions to medicine and science endure. Ajax Carlson the famous physiologist who rose from Scandinavian shepherd boy to unforgettable teacher of physiology was one. He occasionally opened his class in general physiology for nonscientists by sitting down on the platform behind his huge desk, unmanageable eyebrows thatching his penetrating blue eyes, and a gruff Swedish accent rumbling a good morning. Then reaching for one of two large beakers filled with yellowish fluid, he said "von of these is sugar and vater and von is urine." As he took a huge swallow total silence fell over the classroom. Then he would smile and drain the glass—"Ah dis is sugar and water." From then on to the end of the semester class attention was assured. From Dr. Carlson and many more I learned facts and attitudes; ethics, humanity, and serendipity; and the humility to say "I don't know, but I will look it up, and we will then both learn something new."

I was fortunate, perhaps because of my age and my wider experience, to find the clinical clerkships a period of discovery and fulfillment. I recognized that clinical medicine was what I was fitted for, which helped me do well and learn quickly. (Incidentally, my whole class benefited from visiting and observing our girls as they obligingly came down with the succession of contagious diseases such as measles and chicken pox. We all learned firsthand, and the girls were cheered, flattered, and embarrassed by the attention.)

At the same time as our clerkships began, we were placed on active duty in uniform, with pay commensurate with the rank of ensign. This included free tuition, which was indeed a blessing for Connie and me as we had been scraping the bottom of our pot of gold. All our savings from the sale of Ski Lifts and nearly all the funds from the sale of the house not already committed to the bare maintenance of the Calverts and the lake projects were nearly exhausted. Now, with the help from the navy, we were assured of completing my medical training.

We were expected to drill on the Midway, the broad avenue in front of the University of Chicago, very early each morning and to attend lectures on naval organization and etiquette, but it was not possible for me to attend these training sessions from the distance of Seventy-second Street. Accordingly, we moved again, this time to a quite spacious second-floor apartment in an old brick complex on Fifty-third Street. The three-story buildings were arranged in a square around a concrete courtyard down to which wooden back stairs led from the apartment kitchens. Next door was a vacant lot where the crumbling foundations marked the site of a torn-down or burned-out building whose demise must have been long ago, for a magnificent cottonwood tree had grown up between the footings, shading the whole lot. Our apartment fronted on Fifty-third Street, while its entire west side faced the vacant lot.

We were the only Gentiles in a Jewish community that stretched for many blocks in all directions. When we moved in we were politely but curiously scrutinized. For us it was a new experience to be outsiders, and I am sure we committed unwitting gaucheries.

The somewhat stiff but polite barrier that seemed to stand between us and our neighbors was finally and spectacularly broken one wild October evening when wind and thunderstorms swept in off Lake Michigan. The gale battered our building and rattled the windows and tore branches from the cottonwood tree. Lightning flickered, flashed, and thunder crashed and rumbled. Finally the huge cottonwood, half stripped of leaves and branches, gave up the fight and fell thunderously across the lot, its base forming a Medusa head of tangled roots washed clean with the torrential rain. By morning the storm was over and the sky clear and cool. I returned early from work that day and wandered through the lot, climbing over the thirty-six-inch tree trunk and moving the large limbs to one side. Then I fetched my double-bitted ax and starting limbing the trunk, smoothing the jagged, broken stubs of branches and tidying up in general. It was a good feeling to swing an ax again. As is usual with cottonwoods, there were also many dead, dry limbs that had broken off when the tree fell, and I decided to build a fire so we could have a cookout that evening while sitting on the smooth tree trunk. Connie and the children were excited at the prospect. By dusk we had a fine bonfire blazing, hot dogs, and a big pot of coffee boiling on a small cook fire contained by broken cement blocks from the foundation. It was like camping in the big woods. The sound of the girls' singing and the aromatic odor of dry cottonwood drifted with the smoke into the courtyard and down the block. We were no longer worried about the legality of our bonfire when we noticed windows opening all over the building and faces peering out. Then, one family at a time, our neighbors appeared at the edge of the circle of flickering light and asked if they might join us. We were surprised and delighted. Some brought beer and wine and wonderful kosher beef and salami. Most of them—certainly the children—had never seen such a campfire, and they loved it. We all talked and sang and laughed. We made dates to visit, to have cups of coffee together, to walk to school. A police car stopped and the officers came over, uncertain of what, if anything, they should do about our fire. Then they, too, were caught up in the spirit of the occasion and accepted coffee and snacks. The barrier was truly broken that night. By the next morning we exchanged greetings across back porches in a new spirit of understanding and camaraderie.

The more certain I became that I would successfully complete my medical education, the more we all began to plan our future. I had already been offered an internship at Harborview Hospital in Seattle and been granted permission by the navy to accept the position, going on inactive duty for the nine months, which was all that was permitted. Of course we were well aware that the war was far from over. I would return to active duty at the conclusion of the internship, and there would then be residencies ahead before we finally settled down. We decided that when the time came the wisest course would be to work out some sort of loan and build a house somewhere in the Lakes District of Tacoma. Meanwhile we amused ourselves drawing up plans, and I started on a project designed to give them substance by carving a mantlepiece for our ephemeral future home.

Connie telephoned lumberyards, art supply stores, and hobby shops all over Chicago to find the proper wood. She finally located a beautiful clear white pine board two inches thick, twelve wide, and fourteen feet long, which ended up undented and pristine lying on the living room carpet against a wall. The next step was to decide what to carve. The theme that entered my mind was animals—a frieze, perhaps. Connie and the children liked this idea. We discussed what animals would be most suitable, and again we all turned to our

Working on the mantlepiece for the home in the Northwest helped pass the time at medical school in Chicago.

old friend, Ernest Thompson Seton's stories of the proud and beautiful big game animals of North America that were at the same time our neighbors in the Pacific Northwest. It was decided.

During the many succeeding evenings, when I had studied until the words began to run together, I began to lay out the overall design, which I then carved in a deep bas-relief. At the center was an impression of the southwest view of Mount Rainier, while at each end of the carving stylized rocky cliffs would form parentheses. Starting just in front of these cliffs and facing toward Rainier, I drew a mountain goat and at the opposite end a bighorn sheep. Then closer to the center I placed a bounding white-tailed deer balanced by a high-stepping caribou, and next, each facing the mountain, a massive bison bull and a bull moose. The figures were drawn to occupy almost the full width of the plank.

I had whittled a little in the past, never really carved. But somehow I felt sure that I could do justice to this theme. I knew the anatomy of these particular animals, had observed them all in the wild, and felt some sort of empathy and kinship with them since I, too, was at home and content in the same remote regions where they dwelt. Among the few "extras" we had brought with us were my grandfather's wood-carving tools, consisting of a small rawhide mallet, a knife, some gouges, and four other special chisels, all of which fitted into a small rectangular wooden box. I sharpened them and set to work, starting with the sheep and improving my skills as I progressed. It was fascinating and relaxing work. One might even be tempted to say that these tools first used by my Admiralty carver forebears helped to guide my hands. In any case, another Hellyer was using them to good effect and week by week the frieze took form. I worked at night until the tolerant neighbors below tapped on their ceiling, the prearranged signal that it was time to stop the noise. From where I now sit forty years later in the Horseshoe Lake cabin, I face that mantelpiece cut down to fit the big corner fireplace. It is stained dark brown and has gained a fine patina from woodsmoke, wax, and exploring hands.

The regulations about pets in our lease were as strict as before, but this time we ignored them in a grander manner and acquired a number of exotic animal pets, but I will only mention two of them. The first was "Nandi" (*Nandinia binotata*), an African civet. My friend, the small mammal curator at the Brookfield Zoo, gave her to us when she was about ten days old because her mother would not care for her. We took her home, curled up in a basket about six inches in diameter, where she remained hidden under an old wool sock, but if disturbed, snarled and hissed in a most antisocial manner. She was a beautiful little thing. Her long furred tail had faint ringlike markings, dark brown like the rest of her coat but lacking the lighter-colored rosettes that were scattered symmetrically over the rest of her body. She was an agile climber from the beginning, and her sharp claws were wicked weapons as she indulged in temper tantrums during the early months until she learned to sheathe them. Although a civet, she had no disagreeable odor and always urinated and defecated in a sandbox placed in one corner. As she grew, she became more and more affectionate, curling around my neck like a muff while I studied. She had the disconcerting habit, however, of leaping onto my shoulder from a curtain rod or the top of an open door, or even from a wide picture frame that she learned to balance on before jumping. When full-grown she weighed about six pounds and thus landed with a considerable impact. Nandi was omnivorous, but showed a preference for all types of fruit, although she usually shared our meals.

Our neighbors on the same floor were two wispy, timid little spinsters, who always smiled and ducked their heads when they saw us. They used their screen porch, which abutted ours, quite frequently, and often sipped their "medicinal" Mogen David wine there in the evening, becoming a little tiddly in the process, but all in the interest of fending off anemia and unspecified nutritional deficiencies. One late afternoon Nandi managed to slip out onto our porch and was spread-eagle on the screen, not six feet from where the sisters were sitting. One of them suddenly glanced up and saw her and quickly looked away while her accustomed evening flush faded from her cheeks. She did not comment to her sister, who a moment later also noticed Nandi clinging to the mesh. She paled, looked at her wine glass, and the two ladies rose quickly without comment or another glance and disappeared into their apartment. I do not know whether in consequence they resolved independently to reduce the evening wine ration, but they never mentioned their "hallucinations" to us, or anyone else, so far as I know.

As the time neared for our return to the Northwest for my internship, we worried about what to do with Nandi. The problem solved itself, however, when she disappeared one day through an open window. Several days later Nandi's portrait appeared on the front page of the second section of the Chicago *Tribune.* She was sitting on the blue-uniformed shoulder of a Chicago cop, who was smiling happily. In the accompanying story he explained that she was a ferret and had been discovered contentedly crouching on a large bunch of bananas in a small neighborhood store and had now adopted him.

One of the other animals that shared our apartment was a full-grown, irritable raccoon I had found in a pet store, shortly after Nandi's adoption. He looked miserable confined in his small cage. I was alone in the apartment as Connie and the children had returned to Tacoma to find a temporary place to live during my internship in Seattle and stint in the navy.

I was sorry for myself and felt a kinship with that raccoon, rationalizing that I could keep it for awhile, restore it to better health, and then release it out near Riverside in one of the prairie groves before leaving for the West. The pet store man seemed relieved to get rid of the animal and almost gave it away. From the start he was suspicious and sullen. I kept him confined during the day, then loosed him in the evening while I studied. He gradually began to come to me when I sat very still. He would untie my shoelaces and feel the hair on my legs by reaching up my pants legs. If I moved abruptly, however, he growled, humped his back and dropped his head in the characteristic defense-aggression raccoon pose.

One evening when he seemed more friendly than normal and I more on edge as I was in the midst of final exams, I lowered my hand as he was exploring my leg, thinking he might tolerate a gentle stroking. He did not, but instead delivered a savage slash with his teeth. I rose in my fury and pursued him down the long hallway to the screened kitchen door that led to the wooden staircase and the concrete courtyard. The screen door delayed him long enough for me to catch him with one of my best slippered soccer kicks, and when I last saw him he was sailing high over the railing. Immediately I felt remorse. Certainly he must be injured or dead from the fall, and taking my flashlight I searched the whole area below. No sign of a raccoon. I justified my actions as best I could and more or less forgot about it by the next day.

Exactly ten evenings later, same time, same station, I heard the screen door swing open, then slam gently. I looked up, and here came the coon, just as though nothing had happened. I have no idea where he had spent those ten days, but he looked fat and sleek. That was enough. I found a sturdy apple box in the basement, baited it, then waited for him to enter while I held a string that when pulled, closed the top. I finally got him, and when the box was carefully wrapped and tied I was ready to start off. I waited until nearly 2:00 A.M., then, carrying the heavy box by a makeshift handle, I began the long walk through the semidark streets to Washington Park. I did not want to be questioned by a patrolman making his late rounds, for it seemed unlikely that my mission would be understood. Finally I reached the deserted park undetected, passing through the bushes and behind the trees until I reached the artificial pond and stream that occupies its center. There, after a guilty survey of the surroundings, I opened my package and stepped back. Slowly the coon emerged into the half-light, ignoring me. He walked to the water's edge, then, in raccoons fashion, began to feel and search the crannies between the stones that lined the pond with his black sensitive hands, while looking in another direction. Satisfied, I left hurriedly, depositing the box beside the nearest trash can, and reached home shortly thereafter for a very short night's sleep. I do not know what happened to that raccoon, but it is a big park with lots of garbage containers, and raccoons have learned better than any other wild carnivore to live comfortably within the limits of our cities.

Having successfully passed my final examinations, and not having invited relatives to my graduation exercises, I enjoyed the banquets and speeches, feeling free to leave when bored. As class president, I delivered the graduation address. Once these ceremonies were over, I returned home.

The next day I finished packing and was on my way to Tacoma by train. My diploma was in my suitcase to prove at last that I had earned the right to the title of "Doctor."

Internship

ALMOST AS SOON as Connie arrived in Tacoma she rented, for a very reasonable sum, a small cottage. It was charming, although drafty and inconvenient in many ways. Both interior and exterior walls were of rough board and batten. The only heat was furnished by a cobblestone fireplace in the living room and a massive, cranky old wood and coal range in the kitchen, which Connie soon mastered as a result of experience gained at the lake. We had to keep this range burning much of the time as the coils in its firebox heated the water for the bathroom and kitchen sink. On warm summer days a deep but muffled hum and the delightful fragrance of beeswax and honey emanated from the outside living room wall. Even in the winter, when the fireplace was burning hotly, a drowsy murmurous sound and faint scent could be detected from the same source where a large hive of bees, semitorpid after their frantic spring and summer foraging, stirred in their warming wax castle between the studs.

Connie had discovered that she could take a rural bus from the cottage to a small, local grocery store. Despite this, it was inconvenient for her not to have a car while I was in Seattle. Perhaps more out of eccentricity than practicality I bought a used buggy, together with a gentle and very used old horse, as an alternate means of transportation. They were in retirement at a livery stable nearby and obviously not in great demand so the price was right. Prince, the gentle

The horse and buggy were reliable transportation to and from favorite picnic spots.

horse, had a good steady pacing gait and tolerated children well enough to permit all sorts of liberties. There was a fenced pasture next to the cottage, which we were allowed to use, and as all the surrounding roads with the exception of Gravelly Lake Drive were unpaved, conditions were nearly ideal for buggy riding. The idea may have been marginally good in theory, but in practice, Connie and the children never quite got the hang of hitching up. Yet the girls did enjoy taking care of the horse and riding it around the property. When I was home for the two days out of every ten, we hitched up the buggy and drove to various favorite picnic spots, or simply went for a brisk drive—the wheels crunching on gravel, the coachwork squeaking and rattling, and the shod hoofs pounding in the two-beat rhythm of the pace. That fall the teenage daughter of neighbors volunteered to drive the buggy to the lake, and she and her small brother spent a long and adventurous day before arriving wearily at the Calverts. There the horse and buggy stayed. Little used they gradually disintegrated, the horse the victim of abundant food, obesity, and old age, the buggy of mice and mold.

Harborview is the large general hospital serving King County and the entire Seattle area. During the war it was swamped by the multiple demands of war industries—shipyards and aircraft—the burgeoning general population, and the more colorful needs of an expanding Skid Road. The usual quota of interns had been twenty-four, but due to the war there were only twelve of us plus four residents who were expected to staff the facility. It was a hectic time. There were no eight-hour days, our salary was thirty-five dollars a month plus room, and what passed for food in those days of rationing. We found that pepper added interest to mashed potatoes capped with white margarine, for at that time, since Washington supported a powerful dairy industry, it was illegal to mask margarine by adding yellow vegetable coloring.

Despite these minor difficulties, I enjoyed my internship from beginning to end. We were so pushed and frantically busy that we found ourselves performing procedures and undertaking surgeries for which we were not prepared, but we quickly read up on them, looked at the pictures and diagrams, and did our best. It was surprising how little harm we did compared to our successes.

Emergency was my favorite duty—twelve hours on, twelve hours off. Weekend night rotation flew by as the constant stream of emergencies—stabbings, slashings, drunks, and psychotics—poured in. No orderly or city policeman stood by to help with the disoriented or violent cases—only the intern and two nurses.

I can think of no better way to learn and gain confidence than by *doing* when there is no one else available more experienced and more competent than oneself. If I had had illusions about the basic goodness of my fellowman, they were dispelled; yet if my impression of human nature had been jaundiced, I would have found abundant evidence of courage and humor, as well as of enduring frailty. I was impressed for the most part by the steadfastness of the humanity that was carried, staggered, or escorted through those emergency room doors. I was made aware that the deadbeats had not all always been so, and that the drunks, the prostitutes, the paranoids, the manics, and the failed, all had known better times and a very few with help might even find their way back again.

When the children's summer vacation started, we closed the house by simply shutting the door and moved to the lake. By luck the forty acres we needed to protect the west side of the lake from public access finally came up for sale that summer at a tax auction. I attended with needless trepidation, for mine was the only bid—sixty dollars for the forty acres that included the lower homestead. This assured us all the privacy and protection we felt we would ever need.

There had been many changes at the lake during those years in medical school. The caretaker's house was complete. Surrounding it were flower beds and a large, productive vegetable garden, all dug and planted without help by Mary Calvert. The big cedar barn was also finished and a concrete slab and stanchions for five cows gave a look of efficiency to one end of the structure, while the little shed from which the Calverts had now moved assumed its role as milk house with cooler and cream separator. Electricity had been brought in from Clear Lake as far as the house and barn, although it was to be many years before we at the cabin had electric power. Incredibly, Fred had even completed fencing the property and now we were on the verge of becoming a true stump farm.

Meanwhile, we were making ourselves increasingly more comfortable at the cabin. We covered the cold concrete floor with woven hemp squares, and installed a small gasoline water pump in the bathroom, although I often missed the more reliable hand pump by the lakeside, for small gasoline motors and I have never been the best of friends. The times we spent together at the lake that June of 1945 were among the best of our lives. The children loved the life. We had animals all around us—two elkhounds, Linda, and a new young male, Naakve, and I brought home chickens and turkeys and a baby goat we promptly christened Nanette.

Nanette was a French Alpine marked with fawn and dark patterns, an elegant tail, and marvelous yellowish evil horizontally slit pupils. I had bought her from a goat dairy on the way back from Seattle, making sure that she had not been permitted to suck except for cholostrum the first day after birth. I started her drinking from a pan and she learned almost at once, never butting as goats do that have nursed or been bottle fed. She was accepted by the dogs without any difficulties and soon became so imprinted on them that she played with them like a puppy, rambling with them and dashing off into the brush after them when they chased rabbits and other game. She accompanied Connie on the mile-long walk to the mailbox and became housebroken quickly. She found the most convenient entry and exit to be by way of the kitchen sink and window, then out onto the woodpile. In the evening she preferred to lie as close to the fire as possible, her hair becoming singed and yellow from the heat, her eyes half closed in mesmeric, caprid contentment. Young goats are marvelous pets if you have no garden to protect. They are more intelligent, I think, than any other ungulate species and exhibit such a joy of life, impudent humor, and true attachment to people. Nanette loved to startle the chickens and turkeys by leaping out at them, then standing back to watch their indignant squawking and flapping with an expression on her face that could only be called a wicked grin.

Living was quite inexpensive at the lake, since we had our own milk and butter, baked our bread, and had a fairly steady stream of warm brown eggs snatched by Doro from under indignant hens at the moment of laying. She was very patient about waiting for a hen on the nest, although a few eggs were broken as she dashed for the kitchen with her prize, still moist, clutched in her hand.

During this period we saw few of our family or friends

either at the lake or on our trips to town, for most of them were scattered around the world in the military. Brother George was somewhere in the Far East; we did not know then that he was fighting the nastiest of wars in the jungles of Burma, sometimes behind enemy lines—dangerous and dirty work for which he was decorated by the British with the Military Cross. Tay, Connie's younger brother, while piloting his B-17 over Germany, was shot down and declared missing. The story of his time in the underground while being moved through Holland and Belgium and his final reunion with the advancing American troops is one of great courage and a tribute to those civilians in the underground who risked their lives to protect downed Allied aviators and escaped prisoners. Connie's older brother, Larry, had also had a tough war as a naval officer in the Pacific. Even "Uncle Bob" Randolph came out of retirement, going on active duty as the colonel in charge of the Port of Embarkation in Seattle, where he and my grandmother lived during the latter part of the war while we were in Chicago.

Meanwhile my far less heroic contribution to the war effort as a civilian intern fighting the home front battle of acute hospital understaffing, shortages of medical supplies, and poorly trained personnel was coming to an end. I was aware that I had learned a lot about handling emergencies, giving care even when help was inadequate and perhaps most of all knowing when to leave well enough alone. It was now time for me to get back in uniform.

Navy Doctor

On July 1, 1945, my internship and my lake idyll ended and I was again on active duty as a medical officer with orders to report to Farragut, Idaho, for six weeks of special training in the naval hospital, before shipping out. Farragut was a beautiful base on the shore of Lake Pend Oreille where Connie was able to come visit me for one weekend before August 2, when I received orders to report to San Francisco.

On August 6 the bomb fell on Hiroshima.

On August 9 the second bomb fell on Nagasaki. No one was sure what would happen next, but the feeling was strong that as a result of these bombings the war would soon come to an end without the need to invade the Japanese Islands and face the mass horror and unthinkable loss of life that that implied. Although it was a shocking event, at the time relief was uppermost in many people's minds. Feelings of doubt and guilt, which cast dark shadows over the rejoicing, came later with the realization that a new instrument of terror and of great magnitude had been released, the final consequences of which were unknowable.

Looking back to 1942, when almost daily I had passed the secured gateways and walls of Stagg Field on my way to and from classes at the University of Chicago, I had been unaware that in the nether regions of that stadium dedicated to sports and sportsmanship, the brilliant Enrico Fermi and his colleagues were creating the first self-sustaining chain reaction in uranium that led to the development of those bombs.

I arrived in San Francisco on August 9 and reported to headquarters just as the news of the Nagasaki bomb came through. Five days later, on August 14, the Japanese surrendered, San Francisco was pandemonium, and I spent the entire night patching up injured sailors and other celebrants who were brought to the dispensary.

It was an anticlimax to be flown to Pearl Harbor the following day where I was required to report in daily, and then a few days later, was mysteriously ordered back to San Francisco. During the following few months I had many short assignments. Twice I returned to Hawaii only to be sent back to the West Coast, serving temporarily on four different destroyers for short tours that lasted just a few days on two occasions. I traveled by air and by sea, returning the second time on the USS *Saratoga.* On reaching the West Coast I was assigned to the light cruiser, *Mobile,* somewhere off Japan. On October 16 I was on my way again over the Pacific in a C-54 headed for Guam via Pearl Harbor, Johnson Island, and Kwajalien Atoll. We arrived in Okinawa thirty-three hours after leaving Hawaii. After sweeping twice over the muddy landing field, we finally wallowed down in a torrential rainstorm.

The whole area was a vast field of mud and wreckage: beached and sunken ships, rubble, ruined equipment, and more mud. The mud was tan colored, oozy, viscous, so that great balls of it stuck to our feet and gear. Behind the mud flats the green hills were topped with small clumps of miniature palms, and the terraced fields were pockmarked with shell holes and honeycombed with ancient tombs. We improvised quarters in tents that had been collapsed by the wind. C-rations were the only food available. But worst of all, the *Mobile,* which was supposed to be in the harbor, was nowhere to be found. We wondered how long we would be stuck there. Everything was in upheaval from the two typhoons that had hit the island, and the men showed little interest in trying to set things aright.

We stayed there for two weeks amidst rumors that transportation and food were on their way. There was not much else to stave off our boredom as we huddled in our tents listening to the constant rain.

At this time a landing ship (LST) entered the harbor and sent a boat ashore. I learned that the ship was headed for Nagasaki where the *Mobile* might be found. Without any authority or orders I boarded her and was assigned quarters. An epidemic of dysentery kept us from stopping at Nagasaki. We were diverted to the nearby port of Sasebo. Again without orders, I disembarked and caught a train to Nagasaki. The *Mobile* was not there and did not arrive in port until five days later.

While waiting for the *Mobile,* I joined the members of an army medical team assigned to study the effects of the bomb on the civilian population just three months after it was dropped. We first explored the area of the impact, where everything was totally lifeless and devastated, glass and metal fused into unrecognizable lumps and masses. Because the city is divided by a hill, the waterfront and the more residential areas were largely spared, whereas the inland industrial portions received the brunt of the attack. We were all puzzled by the attitude of the Japanese doctors who were represented on our team, for in their eagerness to study the effect of radiation, they seemed more impressed by the enormity and scientific wonder of the destructive power of the bomb than by the human tragedy and the plight of the victims, most of whom suffered burns that ranged from severe to the most superficial flash burns in which the dark patterns of the clothing—a design of a flower or a character—was transferred to the skin like a stencil whereas the light colored background reflected the flash and spared the skin. The full effects of radiation were becoming increasingly apparent, and we observed with horror the victims crowded into every possible facility. In the evenings the Japanese physicians invited us to join them for sake and dinner. Strange indeed are the reactions of recent enemies, colleagues, and oneself.

When the *Mobile* arrived in port, I immediately reported aboard and was welcomed with particular warmth by the doctor whom I was to relieve. We sailed for Sasebo, which had not been damaged by the fighting, and anchored in its filthy, littered bay where dysentery was now rampant. Because of the epidemic, only those engaged in official duties were permitted ashore, and most of us in the medical department waited for the first cases to appear on our ship. This happened despite all precautions two days after our arrival. We worked frantically to contain it, quarantining all heads, sterilizing everything, inspecting the galley personnel for cleanliness, swabbing the decks with disinfectants. We placed all the medical personnel on sulfonamides and segregated them during meals, confining them to the sick bay area of the ship—and miracle of miracles, we seemed to be controlling it, which was more than any of the other ships had done. It was not so much that the illness was serious, although many of the patients became dehydrated and bottles of IV fluids were suspended everywhere from the pipes beneath the overhead in our expanded sick bay, but that if the epidemic became rampant, the ship might be immobilized at sea or doomed to remain in this pestilent harbor for lack of personnel to carry out the necessary duties.

A week later, however, we put to sea with a few of the crew still sick and many very weak, but morale rose visibly as we left Sasebo astern.

My visit to Japan had been something of a fiasco as far as seeing the country. My only real view was on the train trip from Sasebo to Nagasaki and the few days there with the medical group. Coming down on the train, however, we passed very slowly and haltingly through fifty miles of backcountry—lovely yellow rice paddies where the crop was being harvested and carefully hung in neat little shocks over racks to dry—the same apparently unaffected rural Japan of past centuries with its smells and colors gave rise to such mixed emotions in me, familiar yet strange feelings for which I found no expression and certainly could not put into words. I was aware that this was the country of my birth and my home in early childhood.

The people were terribly poor and not properly clothed. The men and boys wore sleezy, ill-fitting cotton uniforms, and the women cheap coolie coats and baggy pants. The older children were friendly. They begged for matches, smokes, and candy, while the little ones stood staring, their flat little noses running semipurulent mucous. Many had the usual impetiginous sores on their shaved scalps. The young men and women did not bow or make way. They displayed neither hostility nor friendliness, and I had no idea what they were thinking. I knew nothing about the wisdom of our occupation policies, or even whether we were dealing with a beaten people. Those living in Nagasaki must have known that they were defeated, but they did not react as I would have expected.

Now that we were at sea, I experienced for the first time what it was like to be on a formal ship. All regulations were enforced: musters, reports, inspections. Everything was done by the bugle, and I could not tell one call from another.

Shortly after sailing, we were informed that our destination was San Diego. We were permitted to send this information home. Excitement and morale were noticeably high. I wrote Connie and suggested that she drive to San Diego. The children could stay in San Francisco with her family.

She arrived in San Diego three days before Christmas, just in time to meet the *Mobile* with all the other wives, girl friends, and families of sailors home from the seas.

The town was jammed, but Connie had somehow found a room in a boardinghouse containing a wide, sagging double bed, an unpainted dresser, and three small pictures, one the portrait of a large, brown dog smoking a pipe; one of a collie whose expression reflected ineffable love for a dove perched on its shoulder; and one portraying two kittens dressed in buttons and bows—none of this affected the joys of our reunion.

Our time together proved of short duration, for quite inexplicably the very next day, Christmas Eve, the *Mobile* was ordered to put to sea for Guam to pick up military personnel—a mission called "Magic Carpet" duty—thus cancelling all promised leaves.

A few days out of San Diego we were in the midst of a great storm. The waves broke clear over both weather decks and a slashing seventy-five-mile-an-hour wind turned the air into a stinging gray sheet of foam. No one was permitted on deck. Cooking was impossible, so we ate sandwiches. Some of the crew were desperately seasick and a few required IVs and sedation, but most were simply dizzy and suffered feelings of unreality from the constant motion and lack of sleep. There was also smoldering resentment at the apparent futility of this mission, for which we were so evidently ill-fitted, and the timing of which was simply inexplicable. In order to counter the low morale and sullen and rebellious mood, I broke out a number of bottles of medicinal whiskey on Christmas evening. By adding quantities of canned milk, eggs, vanilla, and powdered sugar, I created an eggnog of fearful potency and plied many of the officers with this brew, thus hoping to protect us all from the wrath of the powers on the bridge. We spread the cheer as widely and discreetly as possible throughout the ship. The next morning I was filled with dread as I mustered my men, but no mention of the celebration was ever made, and I accounted for the depleted alcohol stores by listing them as broken in a storm at sea. Morale may have risen slightly as a result of the party and markedly with the passing of the storm.

We finally arrived in Guam, where there were no troops for us to take home, thus confirming our belief in the futility of the voyage. After a few days we sailed homeward with rumors rife that our destination was Puget Sound and decommissioning.

Gypsy Camp and the Creek

DURING OUR ABORTED LEAVE in San Diego, Connie and I had discussed how and where I should apply for residency training. We agreed that pediatrics was the speciality I had most enjoyed and in which I had seemed to have the greatest aptitude. But we realized that as doctors were being discharged in increasing numbers with the war's end, the competition was fierce for the limited number of choice residencies then available. This would be especially true for those of us who had chosen civilian internship and thus prolonged our time of active duty. In spite of this, it did appear certain that I would be out by the fall of 1946 as I expected to have enough "points" for discharge by midsummer.

Even when evaluating some of the disadvantages of life in Chicago, it seemed to us that the training I had received had been outstanding as compared to that of many of my associates. I decided to write to the Department of Pediatrics at the University of Chicago, making application for a residency to start in the fall of '46 and sent the letters off from Guam. The mailing of these letters marked the turning of my thoughts toward civilian life and beyond the years of residency, to the practice I hoped to set up in Tacoma.

During my internship we had begun to look for a building site for our future home. The need to be within easy driving distance of Tacoma and its hospitals limited our options to the Lakes District or the city itself. But as we all agreed that we were now confirmed country dwellers, we did not investigate city lots.

At about this time, my log-scaling friend, Pat Erskine, had bought ninety acres of land just off Orchard Road, an arterial that led directly into the northwest part of Tacoma, only ten miles away. The upper or east end of this property consisted of both flat and rolling meadow, dotted with Garry oaks and huge sprawling scattered prairie firs. The land then dipped quite abruptly to a creek bed below, forming a hillside clothed in a magnificent forest of tall, clear firs, hemlocks, and buttressed red cedars. Vine maples, big-leaf maples, and dogwoods were interspersed where sufficient western light penetrated the conifers, encouraging as well a spectacular ground cover of salal, huckleberry, and kinnikinnick. The stream that ran the full length of the property was called Steilacoom Creek after the lake from which it flowed. It would have been called a river anywhere but in this land of abundant waters, for it was about thirty feet across where it flowed over rocks and gravel, but in places narrowed and deepened as the hillside and west bank confined it, then again allowed it to widen into broad shallows. The water was sparkling and so clear that one could see small trout dash for cover as our shadows fell across its surface. In 1946 this was a wild area where deer, raccoon, and mink were common residents and Western gray squirrels were still abundant, as were chipmunks and chickarees, while mountain beaver riddled the hillside with their shallow burrows.

Our hunt for property came to an end on our first day's exploration of this creek. We asked Pat if he would sell us a strip of ten acres extending from the road above to the Steilacoom Game Farm and Fish Hatchery fence on the west—the full width of the property. This formed a strip about a half-mile long. We chose a building site on the west bank where there was open space and gently sloping banks on both sides of the stream. A lovely, grassy flat was held in the bend of the creek directly across from where the house would stand with its back to the fence and open hillside.

Three ancient, black cottonwood trees grew at the water's edge, their light bark and pale green leaves contrasting with the darker trunk and foliage of a massive twisted maple standing in the center of the flat. The sounds of running water soothed and lent further enchantment to this already captivating scene.

To fill the long hours of the return journey from Guam, I had borrowed graph paper, ruler, and other instruments of the draftsman's trade from the navigator and started designing our future house. The site was clearly in mind, and the space requirements were dictated by the size and composition of our family. We needed five bedrooms, two bathrooms, a library with many shelves and a fireplace, and a work area with expanses for activities ranging from major carpentry to big sewing projects. It would also incorporate files, storage space, and built-in music components. Then we envisioned a long and sunny passageway with another large fireplace, eight feet wide by four feet high, expanding into a wide dining and music area with piano and an open kitchen separated only by a diagonal counter. All would flow unpartitioned from one area into the other. The floor was to be a concrete slab with imbedded radiant heating panels, the walls up to the second story were to be hollow concrete block with massive timber headers above doors and windows, the ceiling exposed beams. The upstairs would have plywood floors and walls papered to please the occupants. The building would be thirty feet wide by sixty feet long with large plate glass windows all along the front that paralleled the creek, some thirty feet away.

Gradually, I got the plans ready for Connie to look at, and by the time my ship arrived in Bremerton we were able to go over them, making some changes in detail, but agreeing that the concept was ideal. While I remained shipbound, she showed the plans to several contractors and arranged for bank financing, which turned out to be limited to a ten thousand dollar GI loan.

It was at this time that I received my acceptance for a

two-year residency in pediatrics at the University of Chicago. A few days later I read in the papers that the state of Washington basic science and medical licensing examinations were to be given the following week at the University of Washington. I signed up although I had not intended to take them until after my residency and therefore had not prepared. With some trepidation I took the examination, and after the sadistic six-week period of suspense, found that I had passed and was licensed to practice in Washington—another hurdle jumped. It now seemed to Connie and me that our cup was truly running over.

The house plans, however, proved so unconventional, both in design and use of materials, that no contractor was willing to bid on the job. I telephoned Chet McKasson, who helped build the cabin at Horseshoe Lake, and asked him to get in touch with Mr. Michaud, also of cabin fame, and join us at the cottage on the next Saturday night that I could leave the ship. When that time came we bought some jugs of beer, made snacks, and spread my detailed drawings on the long dining room table. Soon Chet, together with brothers Roy and Charlie, showed up, followed by Michaud, preceded by his fat cigar. We had a fine reunion. I then turned to the plans, told them our problem, and said, "Can you build this for ten thousand dollars—no contract, no guarantees required, but just doing the best you can?" I added that Connie and I and the children would be camping at the site and she would be doing most of the painting while I would help with all phases of the construction as pipe bender, laborer, carpenter, and resident genius.

That evening the pencils came out, the beer flowed, criticisms, suggestions, and arguments passed back and forth. Finally, it was Charlie who said, "I think we can do it, Doc." The others agreed, we shook hands and said good night, certain that we were about to build the house on the creek.

Although lumber was readily available, plywood was in short supply, but we were able to get a half freight car load wholesale at the railroad siding. This we unloaded ourselves onto a borrowed flatbed truck to transport to the creek, where we inadvertently piled it onto my only complete house-plan drawings. From that point on we built from memory. During the first month, while footings were being poured, Connie and the children had moved out of the cottage and set up camp just above the building site and beneath a large fir tree surrounded by graceful and protective vine maples. The creek was at our doorstep providing water. We built a temporary outhouse and did our cooking over an open fire kindled in a fireplace made of nonporous and nonexplosive creek rocks. All our lares and penates were piled to form a canvas-covered protective wall at one side of the camp. The children selected their own bedrooms close by in private hollows between clumps of Scotch broom and salal.

It was one of those unusual but glorious Puget Sound summers when it did not rain from early June well into September. We never needed a tent and entertained our friends with ever-changing furniture made of building materials, the design and composition of which altered as components were suddenly required for inclusion in the house.

Meanwhile, I had been relieved of duty on the *Mobile* and transferred to Seattle, where I was assigned to directing mothballing of the medical departments of two Puget Sound type "baby carriers." All the instruments and nonexpendable items had to be accounted for, cleaned, then wrapped in naval jelly, while all expendable items, regardless of their usefulness or value were to be "disposed of" (which meant "deep-sixed") so as not to compete with civilian products. I tried to get permission to give the large quantities of surplus drugs, battle dressings, and so forth, to the county hospitals, but was refused. I did manage to carry off some of the dressings to Lewis Todd's veterinary hospital on the assumption that neither dog nor master would object to green or khaki bandages. The waste was appalling. I suppose, however, when weighed against the abandoned tanks and guns and jeeps on various battlefields, it was a small matter.

Even more important to me was the problem of trying to find nails for our building project on the creek. They were almost unattainable, but nails were considered expendable items in the ships' carpentry shops, and therefore were to be disposed of, not sold. I disposed of them each night as I left the ship, walking down the gangplank, trying to look carefree with shoulders level, while my briefcase bulged with thirty or more pounds of nails dragging at my elbow. When I delivered them at the creek we decided how many pounds of what size should be liberated on my next trip. This was my only larceny.

I was discharged in June and immediately joined the gypsy camp on the creek. We worked long hours and the house took shape, looking ever more settled and suited to its site than we had hoped. I bent pipes for the radiant heating panels in the floor while the plumber welded them. I dug a shallow well in a clump of blue elderberry bushes just up the creek from camp and built a concrete pump house around it. We kept close accounting of our costs, paying wages and bills for supplies weekly and felt we were doing pretty well when the floor was poured, the concrete brick walls were up, and the second story framed.

Summer was passing all too rapidly. As we were planning to drive to Chicago, we had a definite deadline to meet. Good friends, recently married, wanted to rent our house for the two years that we were to be away. This was a fine arrangement for all of us, and while we were gone they did much more than was ever expected with the house. They even started a garden and a lawn.

About a week before we had to leave, when everything was finished except the main upstairs bathroom, we discovered that our funds had run out. A delayed bill for pipe and plumbing had proved our undoing. The next morning we stopped all work, paid the last wages, and tried to borrow a thousand dollars from the bank, but were refused. In order to rent the house, we had to have the bathroom completed. We had no means of getting the money, for we did not want to turn to our parents for help. That evening we had a party in front of a blazing fire in the huge fireplace, with the McKassons, the plumber, and the electrician, draining the whiskey, beer, and gin bottles that had helped to sustain our joint venture, and bid one another nearly tearful good-byes. It was a sad occasion, but we agreed that they would finish the job whenever we wired and told them that we had obtained the money. We assured them that somehow we would manage.

Connie and I slept poorly that night, rose early and packed the car with everything we could carry, including Varga, our new elkhound. We were a little hung over from the previous evening's wake and still desperate about how we were going to solve our financial dilemma. We had come so close to finishing within the ten thousand dollar loan. We were short only a thousand dollars. Had we had a few more days in Tacoma, we felt we could have found the money. However, Connie's brother, Tay, was to be married two days later in San Francisco, and a large and formal wedding was planned. Our youngest, Tirrell, was to be the flower girl, and I was an usher, so our departure could not be postponed.

We were broke in a sense that we had never experienced before, having saved only enough money to reach Chicago, camping along the way. We knew that when we reached my grandmother's and Uncle Bob's we would be cared for and cherished while we recouped, worked out our financial problems, and sought permanent quarters close to the university. The children felt our anxiety and were depressed at leaving the free and exciting camp life along the creek. We drove that first day until nearly midnight. Having picnicked on sandwiches and fruit and failing to find a good camping place, we turned off into an old gravel pit some distance from the road. Here Connie and I spent a cold, hard night on gravel in sleeping bags beside the car while the children used the car seats. At noon the next day we were in the midst of wedding preparations. I had forgotten a white dress shirt and stiff collar to go with my cutaway and somehow had packed two left dress shoes. A neighbor came to the rescue. The wedding and the reception were beautiful and very grand, but our little family—secretly obsessed by our financial crisis—passed dazedly through the occasion, so lavish in contrast to the life we had been leading.

When we reached Riverside the welcome was so warm and Uncle Bob and my grandmother so delighted with the children that after a magnificent dinner, we poured out our woes about the creek house. Uncle Bob asked a few questions concerning our bank loan and our firm arrangements for the lease of the house after its completion, then said, "Why didn't you damn kids tell me you were having problems? How much do you need?" That very night we drew up a note to cover completion of the house and the next morning wired the McKassons to resume work. Our dilemma was solved. We now looked ahead with assurance, knowing that our house would be completed and that the Calverts would maintain the lake property over the next two years. Our next priority was for me to acquire as much knowledge, understanding, and wisdom in the field of pediatrics as I was capable of without having to worry about our home base.

Residency and Al Capone

WE HAD BEEN INVITED to stay in Riverside for at least a week while we looked for housing on the South Side of Chicago, but had not realized how difficult that search would be. Returning GIs flooded the area wanting to take advantage of the educational benefits suddenly available to them. The University of Chicago had built a Quonset hut encampment close to the campus, but this was already filled. After several days of fruitless search, Uncle Bob suggested that if I thought I could commute the thirty miles from Riverside to the university and arrange for some temporary sleeping quarters in the hospital when necessary, we were welcome to move in with them, and work out some equitable arrangement to share expenses and responsibilities.

This was at once a frightening and appealing proposal. Grandmother Gwinnie was at that time nearly eighty, while Uncle Bob was almost twenty years younger, an age difference not so apparent at the time they married when she was still a young woman in her midforties.

She had been the very beautiful, but spoiled daughter of a ship's captain by the name of Alfred Ackerman. He operated on the Great Lakes and also owned some summer hotels and resorts. It was while living in that resort atmosphere that she and her brothers were greatly indulged and extravagantly admired. Gwinnie was a fine horsewoman, an exceptional rifle shot, and a lover of gaiety and the social life.

Her mother, however, my great-grandmother, Jane Adams, a direct descendant of Henry Adams, was of a more conservative stock. As a little boy I remember her well when she was living out her last days with the Randolphs. She was a tiny, immaculate person, in a white lace cap, and although deaf, she had alert, darting eyes and an acute spectator's smile beneath the dominant Adams' parrot nose. She had approved heartily of my grandmother's marriage to my Scottish grandfather, George Alexander MacLean, who though lacking formal university education, was an avid reader, an able and gentle man. He worked for Marshall Field and Company of Chicago and, at that time, was eligible for a partnership. A tragic automobile accident killed him, his son George, and the chauffeur. My mother was injured seriously, and my grandmother was left a young widow bereft.

It was out of the desolation and loneliness of this tragedy that the marriage to Bob Randolph, a contemporary of my mother and at one time her suitor, came about. Their marriage had predictably sent shock waves through the family and the community, but over time the relationship was accepted. Somehow they had made a life together, always in the same house, and only separated when Uncle Bob served overseas as a colonel in World War I. Inevitably the difference in age became increasingly apparent as Grandmother Gwinnie suffered deafness and the infirmities of old age. As a result, Uncle Bob had given up opportunities for wide travel and distant construction projects that had been offered to him as a brilliant civil engineer.

Nonetheless, both in partnership with his father, Isham Randolph, and later on his own, Uncle Bob had been responsible for many major engineering projects in and near Chicago, some of which bear the Randolph name. He was also interested in civic affairs and had held a variety of positions of importance in the community, ranging from director of operations for the Chicago World's Fair of 1936 to the head and only publicly known member and spokesman of the famous "Secret Six," the organization liberally funded by business and community interests to develop the evidence necessary to bring down the Al Capone empire. It was their undercover work that finally led to Capone's conviction and imprisonment for, of all things, income tax evasion.

Brother George and I felt intimately involved during this phase of Uncle Bob's activities as we happened to be in Chicago staying with the Randolphs shortly after the Secret Six had been formed and was beginning to exert painful pressure on gangland. The atmosphere at the Randolphs' had been tense during this period. There were definite house rules about coming home at night and outside lighting, police patrols, and other precautions, amounting nearly to a state of siege.

One evening—it must have been during a Christmas vacation, probably in 1930—the telephone rang and understandably there was a sudden feeling of apprehension. The call was for Uncle Bob from one of Capone's henchmen, suggesting a face-to-face meeting between the "boss" and Colonel Randolph as representative of the Secret Six. Uncle Bob was cool on the phone, agreeing curtly to the proposal. When he hung up, we asked him what had been arranged. He told us that later that night he was to be picked up by a Capone car in front of the house, that no one was to be informed nor the car followed, and that after the meeting he would be returned in the same fashion. We all begged him not to go but he was adamant. About an hour later a dark limousine rolled into the driveway and dimmed its lights. Uncle Bob, who always carried a .32 caliber automatic in a shoulder holster and kept his .45 service revolver by his bed during these dangerous times, went unarmed. We heard the solid "chunk" of the car door closing as it backed out of the driveway and headed at high speed through the village toward the city.

It was a long night of waiting and anxiety. We felt guilty that we had not been able to stop him, but as promised, at

4:00 A.M. a car slid into the driveway, the door opened, and "chunked" again. The car coasted backward and was gone. We opened the door to an Uncle Bob, still cool but looking unusually grim. He told us that he had been taken to a fancy apartment, though he did not know where, since the car windows and the glass separating the driver from the passenger compartment had been shaded. Everyone had been polite. The "boss," fat, gracious, sporting the famous scar on his face, had talked informally before making his simple proposal: if the law left his bootlegging and gambling operations alone, he would guarantee peace, order, and tranquillity in Chicago, bypassing the regular police force. Such a bargain was not acceptable, and Bob told him so. It was a grim moment, but Capone did not argue. It was accepted that now war existed, that compromises were impossible, and that the investigation would continue.

Capone's organization was infiltrated. Stool pigeons were bought. The shady but privileged were investigated and dossiers were built up. The guilty, the questionably involved, and at times the wholly innocent were included in this sub-rosa invasion of privacy. After the Secret Six was dissolved when the Capone mob had been destroyed, all the dossiers were also destroyed to protect the reputations of the many private individuals whose names had turned up in the records, even if only peripherally.

Since those more suspense-filled days, the Randolph household had become quieter with formal routines and servants to help my grandmother care for the house and garden. It was into this placid environment that I brought Connie and our three little girls; ranging in age from six to nine. All of us were accustomed to a much less rigid life-style.

The sudden explosive expansion of the household into a four-generation family proved too much for some of the servants. The cook departed, and the maid decreased her hours to only two days a week. Richard, the gardener/chauffeur, took over the heavier household chores, while Connie took charge of the kitchen and the overall management of our comings and goings, which, of course, meant daily planning sessions with my grandmother.

We all made adjustments, and though they were difficult at first, the rewards were incalculable. A new cheerfulness and gaiety pervaded our extended family. Uncle Bob loved and teased the girls unmercifully, and they learned to take it and love him in return. The girls enjoyed their excellent school and made a group of playmates, who were welcome to come play at the Randolphs' regularly. My grandmother loved all the bright young faces around her. We all gained from the quiet conversations with Gwinnie, for she had acquired much wisdom in the course of her unconventional and difficult life. To her there were no surprises and no censuring when we talked of our own lives and problems. Uncle Bob and I became very close, and often we would take walks in the evening and communicate as if we were contemporaries. He discussed freely his triumphs, frustrations, and neuroses, subjects that often required responses and a depth of understanding that I found difficult to supply.

During the snowy winters, we seldom had an opportunity for quiet family times with the girls. But as soon as the weather improved in the spring, we took picnic suppers every Sunday night to a shelter in one of the beautiful forest preserves nearby. There, in the glow of the campfire, we could talk to our heart's content. Often the flying sparks from the fire were mirrored in the sky above when shooting stars added to the delight of our family time together. During the humid Midwest summers, we often put the children to bed under wet sheets to cool them. By the time the moisture evaporated, they were asleep. This was a time of homesickness for the golden summers at the lake. Tempers tended to fray, tears came more readily. I drew a picture for our family that showed a sled being pulled by each of us and our two elkhounds in a direction labeled "The Creek" and "The Lake." The caption read: "Is Everybody Pulling?"

Somehow, this sketch made us more conscious of the value of our individual efforts to reach our common goal.

The teaching I received during my pediatric residency was of such quality and dedication that I, in turn, was inspired to transmit as much knowledge and skill as I could to the interns and students. And through this, I discovered that I, too, loved to teach. Many new medical techniques were being developed at that time, and many newly recognized, puzzling conditions challenged us. The babies in our elaborate nursery for premature infants were developing retrolental fibroplasia with partial or even total blindness at an alarming rate even though they were receiving what was thought to be the best of care with multiple transfusions and high concentrations of oxygen. We were also beginning to work with radioactive materials mixed with orange juice in paper cups to treat leukemias.

One strange new condition that particularly fascinated me began to appear with increasing frequency. It was called infantile cortical hyperostosis, in which the cortex of an infant's bones suddenly grow dramatically, often creating swellings and distortions, usually of the long bones. I studied this disease in every way I could, discovering for the first time the fascination of research. I did viral studies using laboratory animals, trying to culture and identify some organism, but found no answers. Doctor Mary Sherman, a good friend from medical school, shared my interest in hyperostosis. She, as a fine bone pathologist, worked on our biopsy material as we collaborated together, even after I was established and seeing cases of this disease in my Tacoma practice. Together we published an early and wide-ranging paper on the subject. For some unknown reason this puzzling and dramatic condition is now rarely seen, as though somehow out of style, or as if it was iatrogenic. We also developed simplified techniques for the earliest exchange transfusions for erythroblastosis, caused by an incompatibility of the mother's and infant's blood, involving, most commonly, the Rh factor, while learning to avoid some of its hazards and complications.

The philosophy of infant feeding and child rearing in the pediatric department was pragmatic and open to new concepts. Many of us fed solid foods as early as an infant seemed to accept them, which we found was at almost any age. We brought babies to the tables for meals at about six months, and we were great believers in breast milk when bottles were in fashion.

There were many fine clinicians whom I admired and from whom I learned, but Dr. Douglas Buchanan probably had the greatest influence on me. He was one of the world's outstanding neurologists, particularly gifted in its pediatric aspects. He taught generations of students, interns, and residents, and none of them will forget his lectures which often started with a phrase or theme, looped out into an apparent digression, then subtly returned to the main topic, only to throw out another loop, return and tie the knot with the very sentence with which he had begun. In his level, somewhat monotonous Scot's voice, he brought history into the clinic. Great men were made real by associating them with their speculations and discoveries. Above all, he taught us to see what we looked at and to appreciate the intellectual delights of serendipity. He reintroduced this word at a medical meeting in the forties. It promptly became a household word, often misused but describing a discovery process leading as surely to Newton's recognition of the significance of the falling apple as to Silly Putty.

Early in the fall of 1947, midway during my residency, Connie and I decided to take a hurried trip back to Tacoma to straighten out some of our affairs at the creek and lake and start the groundwork for establishing a medical practice less than a year hence. We took our dogs—Naakve and Varga—shipping them in the baggage car while we and the children traveled in the cheapest car accommodations. Bill and Virginia Woolf, who were now living just below us on the creek, offered us the use of the small apartment above their garage. We had just two weeks but were overjoyed at the prospect of going home. We all tolerated the discomforts of the train trip quite well. The morning of the last day, however, as the train started the long drop from the Cascade mountain pass into the Puget Sound basin, we were met at breakfast with glum looks from the dining car waiters. They said that they were kept awake all night by incessant barking which started as we reached the mountains. Up to this time the dogs had been silent. We were secretly convinced that it was the smells of the fir forest and mountain air that spelled home to them and triggered their excitement.

On our arrival we were delighted to find that all was going well, that our tenants were satisfied and happy in our house on the creek, and that the Calverts were getting along well at the lake. We had great reunions with our friends, many of whom we had not seen since before the war, and were able to arrange for office space in a comfortable and well-situated clinic in a small detached building on K Street conveniently close to the hospitals. I was to be associated with two of the five Tacoma pediatricians, Drs. Everett Nelson and Charles Kemp, who were already well established.

All too soon we started back to Chicago, but instead of being met as expected by Uncle Bob, my cousin Tom Hellyer was waiting for us on the station platform to give us the news that Bob was critically ill. He was at home in an oxygen tent with a diagnosis of a dissecting aneurysm. I could not believe this finding. It was true that he had been developing emphysema, and probably chronic bronchitis, for all his attempts to stop his chain smoking had failed, but in other respects he had seemed vigorous and youthful with no evidence of cardiovascular disease.

When we arrived at the Randolphs', I went straight to his room where he was lying in bed looking white and obviously fearful. He appeared in some way shriveled and lessened, all the sardonic humor gone from his face. When he reached for my hand, it was as though he were grasping for some ray of hope.

Uncle Bob welcomed my decision that he be admitted at once to Billings Hospital at the University of Chicago where I could be near him. After a thorough investigation it was concluded that he did not have a dissecting aneruysm. His emphysema and chronic bronchitis, however, were so severe that it was necessary to change his previously active life-style and move to a dry and more salubrious climate. He came home from the hospital ten days later, no longer in mortal fear, but dependent at first on the presence of an oxygen cylinder at hand and willing to follow orders. Florida was suggested as a suitable place to live, but he and my grandmother felt it was too far from the Northwest and from my mother in Santa Barbara, so they decided on the Ojai Valley in Southern California.

Accordingly, we set about breaking up the household and its more than a half-century of memories and memorabilia: old letters tied with faded ribbons, diaries, photographs of men and women once known and perhaps loved, but now merely anonymous faces. We burned, discarded, and gave away furniture and curios; some Randolph treasures were sent to the family in the South while favorite furniture was shipped off to California. Within a month the Randolphs were on the train going to a new and uncertain future, leaving behind a lifetime of associations. My grandmother seemed dazed and a little uncomprehending as we said good-bye at the station.

Connie and I were left in the half-empty house to arrange for its sale and deal with the disposal of what was left. It did not sell quickly. As time approached for Connie's departure, we worried about leaving it empty, for she planned to drive cross-country in July with the children and dogs, two months before the end of my residency. We hoped in that way to have our house fairly well settled before the children's school began, and Connie could then attend to the furnishing of my office so all would be in readiness when I arrived.

During this two months alone, I spent as much time as possible at the hospital, for living in the indecency of the ravaged Randolph house was lonely and depressing. I occupied myself by trying to discover which of the remaining furniture, books, and other favorite objects I could crowd into the little van for the journey home, since everything remaining in the house had been given to us to dispose of as we saw fit. Bob's square oak chiffonier with the wonderful drawers and hanging spaces would make a fine tool chest. I loaded it with his well-tempered old saws and other vintage tools from the basement workshop. Even when empty, it was heavy, so I had to pack it in situ in the van where it occupied half the load space. Then there were cartons of books—favorites of mine from the Randolph library—twelve volumes of diaries and accounts of western explorations and voyages published by the Lakeside Press in Chicago; a set of the complete works of Robert Louis Stevenson; and a beautiful, eight-volume leather-bound collection of the Spectator Papers published in London in 1827. In went a gold-topped walking cane, then more books. A fine, big mahogany-framed mirror also just fit through the loading doors when tipped on its side. The van body sank lower and lower on its springs, and the headlights pointed higher and higher into the sky. There was just room enough for my sleeping bag, camping gear, and Connie's cello beside me in the front as I tried a trial run down to my cousin Tom Hellyer's house the night before I planned to start west.

Tom took one look at the low-slung body, flattish tires, and wavering motion and said that I must leave at least half of the load, but I was unwilling to give up any of the precious loot. I spent that night with Tom and started out early in the morning, trying to get the feel of my overloaded vehicle. When perhaps twenty miles on my way, I happened to look back through my rearview mirror and caught a glimpse of Tom following me at a discreet distance, sure that I would turn back. I waved cheerfully and dismissively and saw him finally pull off to the side and I was on my own.

The distance between Chicago and Tacoma seemed very long as I could only go forty-five miles an hour or lose control over the weaving vehicle. And any grade called for a shift of gears. Furthermore, I could not drive at night. Even adjusting the lights to shine as far groundward as possible, they still lighted the treetops more effectively than the road and would certainly bring me to the attention of the law. Driving in this fashion was tiring so I made long midday stops, parking on a wide verge and taking out the cello so that I could reach my lunch supplies. Before loading up again, I often amused myself by sitting beside the open car door pretending to play the instrument. This nearly caused minor accidents as the drivers of passing cars craned their necks to observe this eccentric madman apparently lost in his own fantasies. Finally, after seven days, I started up Snoqualmie Pass, stopping frequently for the car to cool, and feeling like emulating the joyous barks of the dogs in the baggage car as I too smelled first sage and pine, then moisture and fir forest and at last, the lowlands and the Sound. I drove down the dirt road through the game farm to our house on the creek with panache, my hand on the horn and pretending unwarranted confidence, almost overturning the poor faithful van at this final hour.

Great was the rejoicing, great was the confusion of voices, each with a tale to tell. And greatest was the realization that at last I was home, marking an end to our restless wanderings and the beginning of my medical career, finally, at the respectable age of thirty-five.

How to Diagnose a Congenital Naturalist

ON A MONDAY MORNING two weeks after coming home to the creek, I found myself sitting behind my new desk in my new office. My shiny standard black telephone stood alert but silent, while next to it lay a pristine note pad and a well-sharpened pencil. My nurse, Ann, who had just graduated from nursing school in New York, looked starchy and handsome in her white uniform and was as prepared and eager as I that first day.

And so we sat those first weeks, reading, waiting, talking, welcoming each new patient as an honored guest, deserving special treatment.

The office opened on a wide hallway leading to three examining rooms, which were closed off by sliding soundproof doors. Two of these rooms were papered with gay patterned cloth, while the third was larger and designed to please older patients. In general this last room served as a right of passage for the many children who looked forward to the time when they would be ushered into it as a recognition of their more mature status. The examining table was six feet long, and the walls were covered with attractive travel posters. During the next twenty-five years, although the posters became shabby and out of date, the children who had grown up with them would not permit me to change the decor. Three pediatricians occupied our attractive one-story colonial building, and we shared a large, comfortable waiting room, a laboratory, and sterilization room. Each physician had his own nurse and carried his own practice quite separately, although we covered for one another during vacations and times off.

It did not take long before my appointment book was full, as several of the obstetricians began referring their newborns to me, especially after I had done the first exchange transfusions on erythroblastotic babies in Tacoma.

All of our friends in this area were from our premedicine days. I tried at first to avoid seeing their children as patients, because I felt there might be a conflict, but this simply did not work in practice. Soon their children became both my patients and my friends, thus consolidating some wonderful relationships, which continued on in many cases to succeeding generations and have become a source of joy and pride.

Soon after starting my practice I volunteered to spend one day a week at the Rainier State School, an institution for the mentally retarded (or mentally deficient, as we perhaps more realistically termed many of them in those days). It was depressing but medically fascinating. Almost every form and entity of human mental and/or physical abnormality was represented there. There was virtually no medical supervision, although the custodial care was both humane and compassionate. Each time I visited a different ward or hall accompanied by the attendant in charge. Soon Barbara Oblinger, a diminutive extraordinarily perceptive and dedicated clinical psychologist from Tacoma, joined me as a volunteer, taking notes, contributing her expertise and sharing the sensation that we were moving in a world of forgotten beings, some incomplete, half-formed victims of metabolism and genetic errors, some representing well-recognized clinical entities, others conditions that I could not name. With Dr. Buchanan I studied extensively and saw examples of many of the abnormalities, which I was able to identify. Others I could only speculate about, and of course, no sophisticated laboratory was available for diagnostic tests.

Nevertheless, Barbara and I started a diagnostic file in the hope that at one time, proper care would become available, and that the newly founded University of Washington School of Medicine would lend its expertise and take an interest in the institution's teaching potential, which it now has.

During the next four years my practice grew to a point where it was awkward to accept new patients, yet I found it difficult to turn down a plea for help. As a result of my fascination with problems, whether purely physical and immediate, or psychological and requiring extensive involvement, I often worked longer hours than my colleagues. I found, as well, that I was slower at my work than my associates, becoming so interested in each visit, that the usual fifteen-minute office call seemed, in most cases, scarcely adequate. I therefore tried to limit myself to no more than fifteen or twenty patients each day, and in this way was able to keep to a schedule that usually assured little waiting for a busy mother and a restless or sick child.

In those days we made many house calls, usually one or two before hospital rounds and office hours, and others later in the day. Phone calls were also commonly received at night, since pediatricians and/or pediatricians' wives had not yet achieved the skills of self-preservation that now deter the after-office hours phone caller. But despite these time-consuming and often inefficient practices, I flourished. I looked forward to each new day, excited by the challenges, warmed and affected by the esteem and friendship of my "families," and constantly fascinated by the "tricks of the trade" that I seemed to discover almost daily. I listened to wise mothers and indulged my particular fascination with finding answers to common enigmas and to mundane questions rarely dealt with in pediatric texts.

Given my own bent I was always on the lookout for congenital naturalists among my young patients and learned to recognize them early. In times past, small boys seemed most inclined toward the slimy or earthy aspects of natural history. But as a pediatrician-naturalist of some experience, I am convinced that, if anything, the modern little girl in jeans can be just as severely afflicted as her male counterpart and will readily seize the liveliest earthworm from a squeamish, male companion, impaling and threading it expertly for him on a fishhook. Perhaps she will have more difficulty convincing her parents that her obsessive interests are acceptable and irreversible, but, judging by the number of female biologists, applicants to veterinary school, wildlife researchers, veterinary assistants, animal breeders, and jockeys, the battle has been won, and we naturalists of both sexes are in this pleasant plight together.

How early can the congenital naturalist be diagnosed? In my experience, it can be confirmed by eighteen months at the latest, but suspected far earlier. Shortly after six months, when caution and even fear begin to manifest themselves in the human infant, the emerging naturalist shows little dread of animals, smiles, and usually vocalizes when he sees them. These children's favorite books deal with animals. They turn endlessly to the pictures, not selecting them by adult standards of cuteness, but rather because they are large or fierce, tiny or strange.

By three or four years of age, interests become more focused as opportunities to make active animal contacts become more frequent. This is the time for the first pet. No earlier. Some sex difference in attitude to animals is usually evident and continues throughout the adolescent years. The dedicated girl naturalist tends to be more nurturing in her approach than does the boy, more patient and more inclined to adopt the waif and the stray. Her patience and inborn sensitivity often make her a better trainer of dog or horse, and the teenage girl and her horse remain in a very special relationship. An adolescent girl, lavishing her affection on her horse, sharing confidences with it, and yet controlling and dominating a creature so much larger than herself, escapes most of the pitfalls of this perilous period. In rural areas, years ago, boys also shared in the horse relationship, but the lure of wheels and engines now usually wins the contest for their devotion.

There are still other differences between the boy and girl naturalist before their interests merge during the academic years. The boys are far more savage in their pursuits. Few girls are obsessed with the chase. They rarely trap, hunt, or mount specimens, but rather observe, preserve, and cherish. Of course, there are fierce little girls and gentle little boys, and all lifelong naturalists have elements of both.

Pediatrics is not like any other medical specialty, for it is a general practice within an age group. It starts with the baby in the womb, and perhaps ends at young adulthood, but frequently continues in an extended family relationship that not only includes the former patient, but the grandparents, the new baby and its parents as well. It is a demanding relationship that thrives best on mutual trust or at least understanding with these three generations, yet sometimes requires the pediatrician to stand alone against them all and for the child, whose advocate he is in the final analysis.

Pediatrics is also the specialty of optimism and opportunity. Pediatricians have every right nowadays to expect nearly all of their patients to outlive them while becoming year by year less frequently ill, cleverer, and better educated. And this occurs even though the pediatrician may do little but encourage and admire. For it is in the nature of living things that life is most precarious in its earliest stages; in the womb there are many risks, then with birth, and thereafter, hazards of trauma and radical adjustments from a cradled aquatic existence to a harsher air-breathing world take their lesser toll.

The first year usually produces illnesses resulting from immature or absent antibodies challenged by a widening and more complex environment, rare but as yet unexplained sudden deaths, and a few disastrous infections and accidents, but in general it is a period of extraordinary vigor and both physical and intellectual development. Thus, stage by stage through the pediatric years, the incidence of illness becomes less and less until our patient, at the peak of his health, often reluctantly transfers to a physician whose role it will become to preside over the remaining years where the slope soon takes a downward turn, as diseases become more chronic and neuroses, tension states, aging, and all the ills of man become commonplace.

Yet the life of pediatricians is not without frustrations and sorrows, for they share these with the family through the weeks, months, and even years of struggle against some forms of genetic disease, malignancies, and severe mental retardation. There is nothing more tragic than the accidental death of a child who has scarcely had a chance to test the wind and try his wings, and so wasteful and unacceptable as the progressive degenerative diseases of childhood.

Quite obviously these comments and reflections result from hindsight. As I engaged the problems of pediatric practice with enthusiasm in those early years I had little time or inclination to analyze trends or philosophize, and before I arrived home on the creek at the end of the day, my mind had usually purged itself of medical matters on the twenty-minute drive from city clutter to wild woods and running water.

Life on Steilacoom Creek

THERE SEEMED TO BE endless projects on and around the creek during those early years. To improve access to our house, we bulldozed a one-way road that wound through the meadow with its lovely great Garry oaks, then over the brow of the hill and down on a long traverse through deep forest to terminate a hundred feet from our house and on the opposite side of the creek. We felled three thirty-inch fir logs to serve as the sturdy underpinnings of a plank bridge. I welcomed the sights and sounds of this improved route home from the office each day.

Although we did plant a lawn between the house and the creek, the banks were left in a natural state, adorned by foxglove, belladonna, and wild yellow iris. The meadow across the creek with its big cottonwoods and maple trees was left untouched except for a pool and little terrace. When lighted at night the scene from the house was truly magical. In the daytime it was wildly beautiful, whether in the warm, flowery summer or the monochromatic winter when the quiet, friendly creek became a roaring torrent gnawing at the bank and threatening the foundations of the bridge. The streambed shifted yearly when some change upstream occurred. A log caught against a bank redirected the current to the opposite side and so propagated its effects endlessly and unpredictably.

After the bridge, we dug a combined swimming hole

and trout pool. The effect was entirely natural except for the low concrete terrace we poured by the shallow outlet where the meadow joined the pool. There we sat, enjoyed outdoor meals, and even answered the ever-present telephone attached to a leaning alder tree. Sometimes the honking of our Toulouse geese interfered with the usual conversations between doctor and patients. All the "creek folk" enjoyed this pool, and when the weather was clement it became the center for most of the neighborhood's social activities.

Later, to give convenient access to this area, I set nine-inch concrete sewer tiles in the creek and filled them with concrete until the pipes became smooth mushroom-topped stepping-stones. They offered a mild challenge at each crossing. Not infrequently a plate of scrambled eggs drifted downstream to feed the crayfish and suckers that lurked in the quiet depths while a dampened food bearer regained footing.

The children, now twelve, eleven, and nine, were wild with delight at the freedom of their new environment, the spaciousness of their own rooms, and the special secret places along the creek and in the woods that they discovered. We all wanted animals—lots of animals. At one time or another there were sheep, goats, rabbits, chickens, pigeons, burros, pack mules, pigs, and horses. I built a post and rail corral that abutted the creek so that clear, cool water was always available.

The first horse I introduced to this pleasant enclosure was a fine sorrel Morgan gelding with a flaxen tail and mane. Although he lacked the dark bay color and was taller and less compact than his famous progenitor, he had inherited the staunch temperament, willingness, and adaptability of the breed. Tiptoe was his name, which well described his lively way of going. He was three years old and scarcely green broke when I bought him, but he was tractable and soon settled down. He learned a fast, flat-footed walk, head level on a loose rein. At a touch of the heel or lift of the reins, he moved smoothly into a rocking horse canter or, if unrestrained, into a fast, driving run from a standing start. He

Tiptoe was a lively stepping sorrel Morgan gelding whose surefootedness made roaming the rough trails and hills a pleasure.

learned to rein as well as any horse I have ever owned and was surefooted and sensible in the hills and on rough trails.

I rode as often as time allowed, and there was beautiful country to ride over close at hand. The game farm kept a lane open for me along the edge of its property so that from my corral I might pass directly through without opening gates and enter the several square miles of prairie that was bounded to the north by the steep canyon of Chambers Creek and to the west by the high bluff above Puget Sound. Although the terrain rolled, broken only by occasional shallow gullies, it was fairly level and dotted with groves of Garry oaks and prairie firs. There were deer, coyotes, and other smaller game, and no buildings anywhere, although do-it-yourself dirt tracks crisscrossed the lands, often ending in secluded and littered hideaways. It was spacious riding country where I seldom met anyone in those early years.

Wildlife was abundant all around us until the middle sixties when housing developments were started on the prairie behind the game farm, and the meadowland next to ours was sold. Black-tailed deer were common on the creek. Muskrat and mink shared our pool while raccoon traveled the water's edge, prying and probing each hollow and cranny. Red fox and coyotes investigated our chicken house from time to time, and our dogs, although wary, would occasionally approach a skunk too closely.

Twice a year during the salmon runs, the shallow water in front of the house was roiled by the great red fish, already showing patches of discoloration and fungus from their upstream battle with the creek. Many of them spawned in these shallows, and as they chased one another and fought their way against the current their scarred bodies were at times half out of the water. The females, plump with eggs, contrasted with the gaunt, sunken-eyed males whose overgrown hooked jaws gave them a savage, desperate appearance. Pairs paused in the gravelly depressions below riffles where the female dug a nest by vigorous ventral and lateral thrashing motions, then deposited her eggs as the male pushed past her, releasing the milt to flow over them. Many pale coral eggs floated downstream, while others were devoured by trout that followed their larger cousins in anticipation of the feast. All the adult salmon, after giving their final strength to this reproductive spasm, swam slowly off to die. Often they were too exhausted to right themselves in the water and were soon carried off by the current, stranded on sandbars, or entangled by vegetation along the banks.

Then came the predators and scavengers; seagulls appeared from the Sound to feast and quarrel. Crows and raccoons joined the cleanup, and dogs had to be confined so that they would not be poisoned by the decomposing carcasses. In a short time, all evidences of foul odor and grotesque appearing carcasses would be gone, and the end came to this Danse Macabre.

The steelhead runs were quite different, however, for these slender, gray-bodied, blunt-nosed fish were in perfect condition as they moved upstream. They did not crowd the water as did the salmon, but dashed in twos and threes from sheltering rocks to hiding places under cutbanks. They were wary of movements from the shore, whereas the salmon paid little attention to anything but their urge to spawn. In the evenings the steelhead often rested beneath a log or overhanging bank. If one had the skill and if the small devil of petty larceny which lodges in some of us gained the upper hand, one could lie quietly on one's belly, and slowly, oh so slowly, submerge a hand, limp-fingered and raccoonlike, to touch the firm side of an eight-pound fish. Sensing the movement, there is a little forward rush of suspicion, a drifting back along the fingertips, like gently stroking waterweeds, as the gills open wide for a swift powerful grasp of thumb and fingers. With good fortune, a fiercely thrashing steelhead lands on the mossy bank.

The naturalist in me found scope, satisfaction, and delight in the sights from every window of our house and from the life along the creek, the surrounding woods and prairies in and over which I rode.

THE FLOW OF THE CREEK as it emerged from Steilacoom Lake was controlled by a floodgate consisting of heavy planks slotted into metal channels. When the fall and winter rains raised the lake level so that it began to impinge on lawns or threaten the docks of the many lake dwellers, some authority—I could never find out who—removed a plank or two, and suddenly a torrent of water poured downstream carrying debris, silt, and sometimes logs. The protesting voices of the few creek dwellers were drowned out, and each year we contended with high water and some destruction.

It was the second winter after we had built our first bridge that someone, not being content with removing one plank at a time from the floodgate, took them all out at once, releasing an ominous, roaring, foam-flecked torrent that carried away our log bridge. By the next day the creek had subsided. The near side of the bridge was tipped and isolated as the stream had cut a new channel around it, and we were again forced to use the game farm road as our only access.

Late the following spring, Charlie McKasson came down to look the situation over, and we decided to build a new and sturdier bridge with concrete abutments reinforced by riprap extending some distance upstream. Still each winter we were prepared to take swift action as current changed, sandbars built up, and banks eroded; however, Charlie's bridge held firm.

It was shortly after the new bridge was in place that we urged my mother and Tom to come for at least part of the year to be near us. Tom, as the years passed, remembered his Tacoma life with increasing nostalgia, and my mother missed her children and grandchildren who were mostly based in the Northwest. At first they envisioned a simple summer cottage. We were anxious to have them close and suggested that they build on our property, high on a hill at the edge of the woods beside the entry road. From this site they would face west with a wide view over the tops of the trees and hear but not see the creek. This seemed to please everyone. Soon a low, attractive little house was under construction with a glass-walled living room facing west, a tiny kitchen, two bedrooms and baths.

By midsummer the house was finished, and the Ripleys were installed. For Tom, a return to the Puget Sound country where he had spent so much of his life was rejuvenating. His cataract-clouded eyes welcomed the diffuse light and overcast skies. The scent of moisture, fallen leaves, fir, and cedar gave him new energy, and he began to write again, this time not about his Vermont boyhood but about the early Tacoma days. My mother, delighted with Tom's newfound enthusiasms, decided that they should make the move permanent. The house was then altered to satisfy their year-round needs. We turned their carport into a living room library with solid shelves along the thirty-foot-long back wall to accommodate their books.

My mother went to Santa Barbara to settle matters there. When she returned, she never mentioned how difficult it had been to break the ties, leaving the lovely house among the huge rocks and gnarled live oaks of Mission Canyon where she had first come with her two young sons, where she had met and remarried and made with Tom a life of true serenity. Their many devoted friends who gathered around the tea table to give and accept all degrees of confidence and affection were irreplaceable. She sold the house and returned within a few weeks to the newly arranged home in the woods, now furnished with the lovely, old familiar furniture and pictures and warmed by the incomparably satisfying patina of the book-lined wall. Sunny in the front room in summers, dark and encompassing in the back room through the long winters, the tea table moved with the seasons. It was again ready every afternoon, although now it was mostly family who came. Our girls on their way back from school stopped and learned to sit quietly, talk, and listen politely. They knew the comfort of having grandparents to confide in who often seemed more sympathetic than mothers and fathers.

Connie and my mother were always friends and established a special relationship built on a mutual love of books and the instinctive understanding that wise and mature women, free of all rivalry, occasionally seem to form. They drank tea together most afternoons, but if Connie was not able to drop by, there were no hurt feelings. Although I was usually quite late, I often stopped. Tom would leave his typewriter, and we would have a late cup from the pot that my mother always made just in case. This was really Tom's only opportunity for "man talk." He enjoyed my accounts of what was going on in the outside world and what those "damn" Democrats were doing to it.

Tom and my mother rarely discussed politics for he was a congenital Vermont Republican who believed in all the basic tenets of the party, yet had the humor to laugh at some of the pomposities and the compassion to recognize that the Horatio Alger story was not universally applicable. He would have been a good middle-of-the-road Democrat by any other name. My mother, on the other hand, always said that she was a registered Republican but had never been able to bring herself to vote for one. In fact, she was a blazing liberal like many others of her privileged background. In most matters she was clearheaded and logical, but on the subject of politics she was emotionally fixed in the New Deal and Camelot. Nevertheless, they both went to the polls at every opportunity, cancelling out each other's votes. When Tom could no longer see to move the right levers in the voting machine, my mother was allowed to go in and register his choices, although she insisted that he pull the lever and thus kept her record clean.

As his vision failed, his doctor advised him to have the cataracts removed, but the operation proved unsuccessful. He developed complications, nearly dying, and so it was decided not to attempt the second eye. How different it might have been with today's techniques. His typing became increasingly erratic thereafter. If he started with his fingers one letter over on the keyboard, a whole page ended up a jumble, but my mother became adept at sorting out the confusion.

In the spring of 1951, two years after the Ripleys had left Santa Barbara, Uncle Bob died suddenly of a coronary thrombosis. My mother and I flew down together to be with my grandmother and arrange for the funeral. Gwinnie was shocked, confused, and quite unable to think for herself, so we decided that we would have to bring her back to Tacoma with us.

Before Uncle Bob's funeral, the young clergyman at All Saints by the Sea, where the service was to be held, sat down with my mother and me. He asked, "What kind of a man was Colonel Randolph? I always like to know something about the deceased so that I may pay an appropriate tribute." We looked at each other, my mother and I—which Bob Randolph—hers or mine? The arrogant, teasing, sardonic, and often bitter man? Or the loyal, patriotic, dedicated and loving stepfather, who had begun as the stern but affectionate head of our extended family household in Riverside and ended after his illness as my close, emotionally vulnerable, and dependent friend? The Colonel Randolph of engineering and civic fame, or the cynical, private man who confided his frustration to the pages of his diary? I answered for us both, telling of his accomplishments and his loyalty, of his religiosity yet gentle mockery typified in his grace before meals, "Lord, make us thankful," and ended the meal with, "Thank God that's done!" There was no doubt in any of our minds that he had loved us all differently and felt in his own way that George and I were the sons he never had. He was a complex man whose half-smiling, world-weary expression greets me from his portrait each time I descend the spiral staircase of our lake house.

The funeral was simple and moving. The coffin was draped with the flag presented by the local legion. His body was shipped back to Hinsdale to lie beside his father and the other members of the Randolph tribe who had moved from the South to make Chicago their home and there leave their lasting mark.

Connie and I, who had been sheltered and loved by Gwinnie for two years in Riverside, decided that she must come to us just as we had gone to her. Accordingly, we moved out of our big bedroom with its dressing room and bath, taking over Doro's quarters, while she doubled up with Connie Anne and Tirrell to make way for a nurse. With her favorite furniture about her, a fire in the fireplace, and the creek and meadow lying just beyond the window wall of her new quarters, she suddenly brightened up and became almost herself again.

It was just before Christmas, and my brother was home from Formosa for the holidays. The salmon were running in the creek, and my grandmother sat at the windows by the hour, watching them dashing about in the shallows. For the first time she felt able to come downstairs for a festive meal with both her grandsons, her daughter, all the in-laws, and the grand- and great-grandchildren. She was overjoyed as I had seldom seen her before. She was just stepping into her favorite evening gown, her hair all carefully dressed when she gave a sudden cry. We heard her fall and by the time we reached her she was gone. We all felt sad, but not grief-stricken. The stroke could hardly have come at a brighter moment for her, with all her family around her, Christmas, and the drama of death and renewal acted out before her fascinated eyes by the spawning of the salmon.

Japan Revisited

In 1956 we had been married twenty years, and to help us celebrate this event, Connie's mother and father gave us a round-trip ticket to any place in the world. It was a stunning offer, which we accepted most gratefully. The question of where to go nearly precipitated our first serious disagreement. We pored through the atlas—Greece? Back to England and the Continent? Africa and its wildlife? We finally settled on Japan and a visit to Aunt Marion Liddell in Hong Kong.

The flight from Seattle to Tokyo took nearly thirty hours, with stops in Anchorage and again on the wind- and rain-swept island of Shemya, near the tip of the Aleutians. When we finally sighted the coast of Japan, the sun was sinking, and the paddy fields reflected light like a mosaic of irregularly shaped cut glass. Unfortunately, we were so worn-out and burdened with a feeling of total unreality, that we did not appreciate this sight fully at the time.

Landing at the Tokyo airport two hours late, we were immediately assailed by crowds of milling people, all talking at once, while only groups of uniformed schoolchildren formed dark islands of order in the frenetic bustle. It was like this everywhere in Tokyo, as if an anthill were being stirred up by a stick in the hands of some mad, mischievous child.

The Old Imperial Hotel where we stayed has long since been replaced, but it was magnificent in its time—among

Frank Lloyd Wright's most memorable buildings. An amazing structure, it had variously been called "The Catacombs," "The Mayan Tombs," and "The Monstrosity." Mayan tombs seemed the most appropriate image, for the hotel was built on several levels of brownish beige volcanic stone, sprouting barbaric knobs and other excrescences. In the public rooms high clerestory windows let in dim light, accenting with shadows the low matching furniture in the Frank Lloyd Wright style. The guest rooms were low ceilinged with beautifully crafted built-in dressers and chest of drawers, all sized to the Japanese of three generations ago, but a little cramped for the European guest.

The next morning, as I was standing at the counter of the Japan Travel Bureau, a slight, distinguished-looking man came in and stood beside me. He was olive-skinned with aquiline features, and dressed in a conservative, dark blue Italian silk suit. He spoke with a New York intonation and might have been anywhere between forty-five and sixty-five years old. Turning to me, he smiled and asked where I was going, and when I said, "Kyoto," he stated that he, too, was headed that way and had to be in Nara to attend the press party celebrating the completion of the filming of *Teahouse of the August Moon.* His name was George Delacorte, owner and president of the Delacorte Press and publisher of movie and other special feature magazines. He was traveling alone with a private car and chaffeur. When we said we were going to Miyanoshita on our way to Kyoto and told him something about its beauty and its boyhood memories for me, he suggested that we join him in his luxurious car and travel there together.

We had a beautiful drive to Odawara, then up the valley over the same cobbled road I traveled as a child. The last of the cherry trees were still in bloom. As we wound up the narrow way, the blossoming trees far below us looked like their own petals scattered among the dark evergreens. Memories flooded back. I had not been here since my fifth birthday, but the lovely old hotel and gardens were as I remem-

The carp pool at Miyanoshita, where Hellyer tea merchants had posed some fifty years previous, was still beautiful and serene in 1956.

bered them—the pools, the carp, the long-tailed roosters, the high-ceilinged comfortable rooms. After a delicious lunch in the ornate, colonnaded dining room with its intricate and delicately carved ceiling and frieze of animals and flowers, we explored the various baths, some like woodland grottoes—all fanciful and luxurious. We climbed the "crow's nest," a three-story cupola, to see the photographs of the famous—movie greats, kings, heads of state, and many more—who had visited the hotel and had consented to be photographed with Mr. Yamaguchi and his incredible sweeping mustachio.

After three lovely days at Miyanoshita, we returned to Odawara, then continued on to Kyoto while George visited Mr. Mikimoto of pearl fame to study pearl culture. We were to meet again in Kyoto where he had invited us, as the only outsiders, to attend the *Teahouse of the August Moon* party and mingle with the movie stars.

Two days later we met George at his hotel and we drove in an MGM car through the lovely spring countryside to the movie set. Nothing had been dismantled. The goats were still in the huts, the airy teahouse intact, and the lovely gardens well tended. On closer examination we could see the camouflaged cables, which would lift off the roof and move away the walls when it came time to disassemble the house. We ate a lavish Japanese meal with the other guests. The manager of the Kabuki theaters in Tokyo took a great shine to Connie, and they carried on a conversation in broken English. We also met and talked with Glenn Ford, who was charming and low-key, and with Marlon Brando, still with Caesar hairstyle, who appeared nervous and unsettled. He was accompanied by his aunt, who gravitated to Connie immediately, explaining that "Bud" was upset because his bongo drums had not yet arrived. Then we went back to the Nara hotel with George. The formal party was under way—geishas with their astonishing white makeup and doll-like gestures, newsmen, producers, directors, the mayors of Kyoto and Nara and their entourages, and of course the actors: Brando, Ford, Eddie Albert, Louis Calhern, and Machikokyo, the lovely, serene Japanese star.

Neither Connie nor I are movie fans, but we tried to see this marvelously colorful happening through the eyes of our daughters, and so took many rolls of film. At last it was over, and we headed back to Kyoto. George decided he would meet us in Hong Kong a week later. We spent a few peaceful days sightseeing in Kyoto. It was lovely and then the least westernized of the major Japanese cities. We stayed in a Japanese inn with our own little house in the midst of a lovely garden. Our maid attended to all our needs. We bathed in our own bathhouse in the Japanese manner and slept with beanbag pillows on futons on the tatami-covered floors. Before returning to the hysteria of Tokyo, we broke the journey in Shizuoka to see if there was anything left to recall my childhood. There was not. I failed to recognize a

George Delacorte invited Connie and David to rub elbows with actors from the film, Tea House of the August Moon.

single landmark except the beautiful view of Mount Fuji and the climbing hills behind the half-modern city where the tea bushes like flocks of green sheep still fed in the thin mountain soil.

We next went on to Hong Kong. We were met by Wong, the chauffeur, who whisked us away to be greeted by Aunt Marion, who at eighty was still beautiful and regal, yet full of humor and vitality. From the balcony of her house, high on the hillside above Repulse Bay, the China Sea lay like a relief map, its surface color and texture changing with the time of day and the cloud patterns in the sky, while countless islands rose above the water like the undulating backs and flattened heads, knees, and elbows of sea serpents, or perhaps Chinese dragons. Sampans and junks moved slowly across the open water, their patched and graceful sails aslant to catch the erratic breezes. Toward evening as the sun set red and the water turned to steel, fishermen blew their conch shells. The hollow owl tones added to the strangeness and beauty of the age-old scene. Yet not so strange, for a world away we knew islands and water—sunsets on Puget Sound, fishermen in sturdy trawlers and purse seiners, pleasure boats, the sound of a horn or a radio playing. They were consonant worlds, one no less impressive than the other, but one made magical by its remoteness, the other made commonplace by its familiarity.

The Hong Kong Aunt Marion arranged for us to see was the nostalgic remnant of a passing way of life. At home we lunched on tiny sparrows—each a delicious bite served by the maid and the elderly "boy." A charming expatriate French woman whom Marion befriended took us shopping. At formal dinner parties we met the different social groups—the diplomatic people, then the old guard, both European and Nationalist Chinese, who had been driven from Shanghai and Peking. A charming Chinese couple entertained us at a four-hour-long Peking dinner. They had fled the Communists, bringing little money with them but had saved their fabulous jewelry and objets d'art. From the sale of these they were able to live in style. The servants who waited on us wore queues and dressed in the old manner. I sat beside the hostess, who, charmingly ignoring my protests, replenished my silver monogrammed plate each time a course passed by. As the Chinese wine circulated everyone looked at me as the guest of honor, and in turn toasted me, while I replied in the traditional way by tossing off the small glassful that was refilled immediately.

As a result, I was in agony as the evening dragged on. My always sensitive stomach revolted at the repeated insults and my head swam. The room in which we sat (there were fourteen of us, half Chinese and half Caucasian) was closed off from the main parts of the house by an iron-barred door to discourage thieves. And although my bladder craved relief, I could not ask to be excused. It was after midnight when we rose from the table and said our good-byes. As Wong drove us up the winding road toward home, I told him to stop. Ignominiously, I crawled over the stone parapet that guarded the roadside, and there lost in his eyes not only an enormous amount of face, but an incredible amount of bird's-nest soup, shark fin soup, fish soup, three first entrées, and three second entrées, among other Chinese delicacies. It was an experience I did not know how to handle graciously—never again.

One day we went to the races and sat in a magnificent private box where drinks and a delicious luncheon were served. We watched the horses run or parade in the paddock directly below us as a servant placed our bets for us. Between each race a line of barefoot Chinese women dressed in long, black sacklike dresses and broad hats with black fringes around the brims tucked the turf back into place with their bare toes, replacing the divots that the galloping horses displaced. I won enough Hong Kong dollars on those races to buy some ivory horses that I had coveted, a fine souvenir of another time and another way of life.

At last our vacation time had come to an end and the pull of home on the creek, the lake, the simpler life, and of my practice, gave way to a real yearning. We said good-bye to Aunt Marion, thanking her and wondering when we might ever see her again. We rode to the airport in a tropical downpour.

In December of that year, Tom Ripley died after a short illness. He was ninety-two. His life was one of joys and fulfillments. A Renaissance man, he played the flute, the Spanish guitar, sang, and loved to dance. He was witty and urbane. He did fine carpentry, creating elegant furniture out of antique woods, made bows and arrows, batiked, painted in oils well enough to exhibit in well-known shows, and wrote countless charming letters and two books, *A Vermont Boyhood* and *Green Timber.* Above all, certainly in the years I knew him, he was a counselor to me, a companion, and my best male friend.

During his last days he could not talk but he knew us all and made plain that only I was to start his morning IVs, and so each day after my hospital rounds, I visited him. He would be waiting with arm extended and his best effort at a smile. The IV was always ready. I slipped the needle through the resistant, ropey walls of his old veins—I who was so used to the tiny, delicate threads and strings of infant vessels—and we communicated wordlessly. One morning, about a week after his stroke, he did not offer his arm, but shook his head almost imperceptibly, for he and all of us knew that his time had come.

His simple funeral was attended by most of his family, his own by blood and his own by marriage. It was not a sad time, for his spirit, gentility, and charm were still there with us. As the service ended with a rousing, "Onward, Christian Soldiers," as he had requested, it was as if he had quietly but firmly closed the door behind him.

My mother was lonely, but she remained self-sufficient with her books at hand and her family nearby. She started a routine of frequent visits to Santa Barbara. Her old friends welcomed her and helped to restore in her the confidence that she could still hold her own in a sophisticated and intellectual atmosphere. It was wonderful for her to be with her contemporaries who remembered a world they conceived to have been gentler and less complex. When my mother felt ready, she returned gladly to the simple life on the hill above the creek and to an atmosphere where teatime visitors were the very young, concerned with school and boys and hairdos, or their parents, concerned with the here and now and the near tomorrows.

Publishing with George Delacorte

IN OCTOBER OF 1963, I reached the impressive age of fifty and felt I had learned something about life and living and my profession. For some time I had been thinking about setting some of my thoughts and pediatric experiences down on paper. Our three daughters, all married and starting families of their own, as well as many of my patients, had been urging me to write a book to help them steer a simple and commonsensical course through the difficult early years of child raising.

Until this time I had felt too involved with my practice, the family, the lake, and a number of civic interests to write a book. Early in the spring of 1964, however, there seemed to be no reason to procrastinate, and Connie reminded me of George Delacorte's offhand remark during a delightful evening we spent together at Miyanoshita—if ever I were to write anything, he would like to see it. I then began tapping out the pages on my small portable.

In the foreword to this book I explained my aims and motivations in the perspective of that time, starting with a quotation from Kipling's *In the Neolithic Age,* "There are nine and sixty ways of constructing tribal lays, And—every—single—one—of—them—is—right." I went on to point out that many of the mothers to whom I talked had read most of the available books on child care and yet had remained confused by the doctor's varied and often ambivalent approaches. They felt a great need for a simple and direct, month by month, period by period account of their child's development with concrete suggestions for meeting these stages and anticipating future events.

I had become convinced that when mothers understood that certain actions and behavior were a direct expression of the level the child had attained—physical, intellectual, or emotional—the mothers would welcome them as signs of normal development and readiness. They were triumphs for the child, giving satisfaction and fulfillment. Frustration indicated that either the parent or something in the environment was preventing this expression or that a temporary imbalance existed between physical capacity and intellectual outreaching.

I had found that a somewhat firm direction and setting of limits, although admittedly difficult to maintain, was one key to successful relationships. The needs of both the parents and the children had to be recognized and honored if the goals for the children were to be the same as those of the parents. This did not imply conformity but a mutual recognition of socially acceptable, nondestructive courses. Such direction moderates not only the rebellious child who scorns lessons of the past and adopts behavior which can close doors to the future, but also the child who might have a tendency to sacrifice adventure and independence for security.

Most basic of all it is in a climate of mutual trust and esteem that children fulfill their intellectual potentials, as well as respect knowledge and those who proffer it. The ethics we attempt to instill should emphasize standards of decency and compassion. They should be able to withstand the emotional and irrational onslaughts of fads and isms. They should be capable of righteous indignation in the face of cruelty and injustice. Yet they should be tolerant of differing interpretations. It is no coincidence that parents who have a capacity for warmth, humor, generosity, and self-discipline produce children with those same qualities. In turn those children understand the joys of accomplishment according to their unique abilities and continue to expect no less of themselves throughout their lives. I closed the introduction with the quotation from Thoreau: "If a man does not keep pace with his companions it is perhaps because he hears a different drummer. Let him step to the music which he hears, however measured or far away." I then covered myself from this bit of liberalism by adding a line from the twenty-second proverb: "Train up a child in the way he should go; and when he is old he will not depart from it."

The format of the book is a series of essays written for my daughters and their contemporaries, discussing the most critical period of a child's life, the time from birth to five years old. It took me about eight months to finish the first draft, with Connie helping me all the way, retyping, editing, and encouraging. When in the fall of 1964 we decided to attend the American Academy of Pediatrics meetings in New York, I wrote to George Delacorte about our plans, reminding him of our long-ago evening in Japan and informing him that Connie and I hoped to see him and incidentally that I was bringing a manuscript along.

After the medical meetings were over I called George, and he invited us to lunch at his club near the Plaza Hotel, where we had a grand reunion enlivened by fine wine. As we walked out into the bright autumn sunlight, George turned toward the southeast entrance of Central Park, where Saint-Gaudens's handsome equestrian statue of General Sherman stands, with its fountain fed by a feeble stream of water. With a wave of his hand encompassing the entire square, he made an amusing anatomical comment about the feeble output of the fountain, then described his plans to turn the area into a "people place." We were somewhat overwhelmed by the breadth and scope of his vision, but before we had time to visualize it completely, he took Connie by the arm and swept us into the park.

As we started north, he often recognized and was greeted by old people sitting on park benches or strolling in the sunshine—mostly men, whose stations in life ranged from the derelict to the impeccably dressed and favored. He seemed to be a familiar and admired figure, who had on occasion shared a park bench with them. When we reached the zoo—a truly urban menagerie but well kept, I was not permitted to linger by the yak enclosure where some very intense and shaggy amatory activity was in progress. Instead we stopped at the arch, on top of which a wondrous clock was being erected. George explained that this clock was one of his gifts to the city and when completed would show animals in motion as bells chimed on the hour. Not satisfied by some detail of the clock's base, he started up a rickety ladder that was left leaning against the arch, for it was Sunday and there were no workmen about. I steadied the ladder, but George was very spry at seventy and came back down with no difficulty.

We proceeded north through the park at an energetic pace, until he paused at the fascinating bronze grouping of Alice in Wonderland and her fellow guests at the Mad Hatter's tea party. As usual, children climbed all over the figures, and Alice's bronze nose shone from the gentle stroking of many small, loving hands. We stood awhile watching this scene, and it was obvious that George was deeply moved, for he had commissioned the work and given it to his beloved city of New York in memory of his wife, Margaret.

Our next stop was at the Delacorte Shakespeare Theater—again a gift to the city—and there George searched for a hidden key to the back entrance so that he might show us the interior of the theater but was unable to find it. As a park policeman began to show interest in our activities, George decided to move on and avoid complicated explanations.

Soon thereafter we reached the north end of the park and his apartment across the street from the Metropolitan Museum. And what an apartment it was—a fountain in the foyer, marble statuary, fine paintings and sculptures by Murillo, Rubens, and other old masters, period rooms furnished with meticulous attention to detail. His lovely new wife greeted us, and we adjourned to one of the less formal rooms for drinks, most welcome after walking the full length of Central Park.

Up to this time, my manuscript had not been mentioned, and I still held the soft briefcase in which I protected it under my arm. The time had come now, if ever, to bring up the subject of my book. Somewhat hesitantly I said, "By the way George, here is the manuscript you asked me to show you." It took him a moment to descend from the high flight of his plans and dreams, and then he said, "Oh, of course, leave it with me and I will pass it on to the editor-in-chief, who will decide whether we can use it; as you know, I am nearly retired now and don't have much to do with publishing." And there we left the matter with my precious manuscript unceremoniously shoved into a drawer that showed little evidence of being opened frequently. We said our warm good-byes, George promised to come and see us the next fall as he was planning a trip West.

We waited and waited. Two months passed, and finally I wrote to Dell inquiring about the fate of my manuscript. I enclosed with my letter a favorable criticism of it by my friend and former professor of pediatrics, Howell Wright at the University of Chicago. Three weeks later I received a note from George's secretary stating that the manuscript was not enclosed with my letter. She also reported that Mr. Delacorte did not know whether I sent the manuscript to someone else at Dell or whether it was accidentally not enclosed. I wired the next day, telling her that Mr. Delacorte had a copy of the manuscript in a drawer at his home, where I had given it to him November 6. On December 17 I received a letter from Ross Claiborne, the editor-in-chief. The manuscript had been passed on to him and had had two enthusiastic readings. He promised to get back to me in January.

It wasn't until February that I heard from him again. He apologized for the delay and assured me they would have news soon. Two days later on February 12, 1965, he offered me a contract. Soon after that, George wrote with his

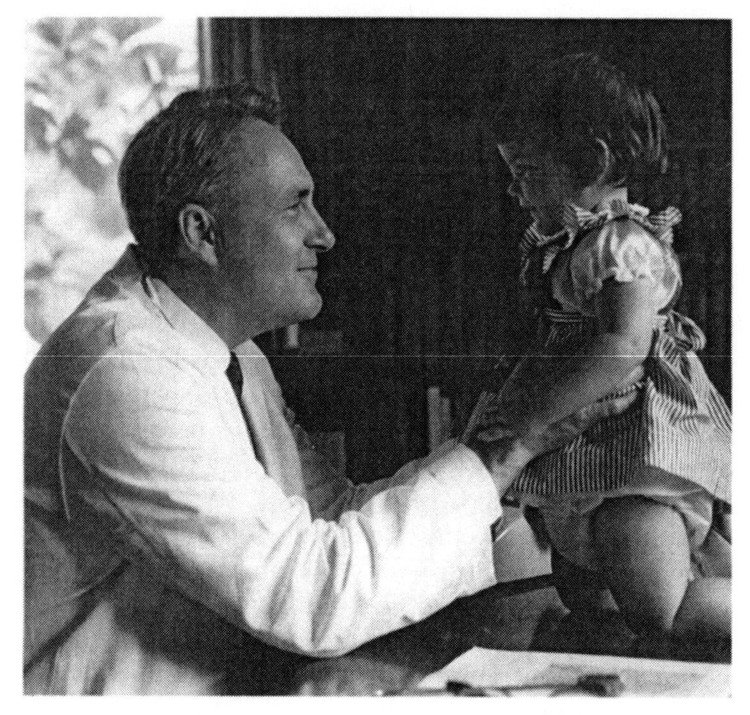

Cover photograph for Your Child and You: A Pediatrician Talks to New Mothers. *(Photograph by Jeannie Dellaccio.)*

congratulations.

When the manuscript came back, there were only minor changes suggested. Someone with a bright red crayon had done little but change nearly every "which" to a "that." Returning the manuscript, I wrote the associate editor, in part, as follows, "The corrections I accept with only minor exceptions. In fact, if we come to New York, I hope you will give us the pleasure of joining us for lunch and cocktails at the rendezvous *THAT* you prefer and *WHICH* serves the best double dry martinis."

At last the galleys came, were gone over and returned, the flat copy was approved, and the cover photograph of me and my first granddaughter, Victoria, "Tory," accepted with enthusiasm. Then, suddenly, at some sales meeting at the last moment, they decided to change the title to an unimaginative *Your Child and You: A Pediatrician Talks to New Mothers.* Portions of the book soon appeared in magazines. The paperback edition came out a year later, went into second printing, and enjoyed a modest success for several years.

The Old Order Changes

ONCE HAVING COMMITTED my beliefs and sometimes unorthodox conclusions to paper, Connie and I were now more free to turn our attention and concerns to the lake property, now Horseshoe Lake Ranch, since we had started raising beef cattle after my discharge from the army in 1956. With heightened awareness we wandered through the maturing forests in their varied beauty: the young hemlocks and cedars just beginning to occupy the understory, and the brighter alder groves carpeted with red huckleberry, blackberry, and abundant plant life, stimulated by the filtered shifting light, especially at the forest's edge. The lake appeared more clear, and the other wetlands more fecund and varied both in plant and animal life.

Our interest continued to shift from the creek where suburbia was rapidly invading the wild or unsettled areas, creating the vast sprawl of Lakewood. The prairie land beyond the game farm had been subdivided and was no longer my riding preserve, and our horses and mules were now at the ranch. Both above us and below us on the creek, new housing was springing up while our upper meadow was no longer usable as pastureland or for any other activity since the property on both sides was being cut up into lots. Soon our access road would be lined with houses.

The big house on the creek empty of children, the large area of lawn and garden, the continuing problems of repairs, irrigation, and household water system, the bridge, annually assaulted by high waters all conspired to make Connie and me consider a major change in our life-style.

It was at this time that my mother's health began to fail. She had begun to have short and evanescent episodes of cerebral anoxia marked by mild confusion and some difficulty finding the words she wanted—she to whom words and the uses of language had always been a chief joy whether in books or conversation.

In the spring of 1968 she had several small strokes, then a more severe one that left her with some aphasia. The morning of June 1, the housekeeper found her on the floor where she had fallen during the night. She was in the hospital for about a week but never spoke again. George was in the Congo as a counselor to our embassy at the time in the midst of dangerous political turmoil and could not be reached even through diplomatic channels. Finally a ham operator by a series of intricate relays got word to him. He flew home, arriving on June 7 and reaching her bedside just before she died. I think she knew we were both with her there.

My mother's death and George's presence prompted us to make decisions quickly. We put her house up for sale at once, and Connie and I agreed that the time had come for us to make a move. It was no longer necessary to have our headquarters in Tacoma now that my mother was gone. We would use the proceeds of the sale of the creek property and house to buy a small, convenient condominium close to my office. With the remainder we would build our permanent and final home at the lake.

But even so, leaving the creek was like turning our backs on a period of our lives, which all of us, parents and children alike, had experienced with nearly pure pleasure. The discovery of the magic site, our camp on the stream bank beneath the vine maples while we built the house, the animals, the "creeknics" with our neighbors, the lovely garden that Connie coaxed from the stony prairie soil—all are cherished memories. There the children grew up, went to school, and returned through those lovely woods. They were married in that house. It was the site of the chamber music concerts where good friends gathered twice a year to hear fine music and enjoy convivial evenings. It was a place where we as a family had a special closeness and played, read, argued, discussed, and established rules and standards, expecting excellence. When the time came, we all grew up and moved away from the creek—the children because it was the moment for dispersal, Connie and I because the area was changing, and we were no longer isolated. There were no dreams there for us anymore.

As planned, we bought a lovely small condominium just two blocks above the waterfront in "Old Town." It has a fine view from the front windows of the Olympics, the islands, and the bay. From the back windows we see the tideflats, port, and industrial complex of Tacoma—ships coming and going, sails, log rafts, all the activities of a busy harbor create an ever-changing scene. For me, the proximity to the office and hospital made my last three years of private practice much more convenient.

We then began to plan our final house at the lake. After considering several approaches, we decided on spaciousness that incorporated the existing cabin, which would be left relatively unchanged in its original Western flavor. The rest of the house, though informal, was to have an open kitchen, a library, and wall space for our favorite paintings. Work started in midsummer of 1969 with the McKassons in charge but operating without contract on a cost-plus basis. We moved in December of that year, just in time to have Christmas in our warm and welcoming new home.

Although 1969 was an eventful year marked by a reorientation of our lives, it was also a sad and portentous one. Not only did my mother die in June, but Connie's parents, who lived in a magnificent house in San Francisco overlooking the Presidio and the bay, were in their eighties and in failing health. Connie's mother and father were extremely close, inseparable and dependant on each other as few couples are. When her father died at home, quietly, of cardiac decompensation in November of that same year, her mother found no further motive for carrying on. She was happy to see her grandchildren and children, and Connie visited often, but her conversation was mostly of the past, and she spoke of Connie's father as though he were a continuing presence. She developed a gradually increasing weakness, and after a short illness, she died in May of 1970.

Shortly after we moved into our new house, Connie and I found ourselves sitting at the head of the table with all of the previous generation gone. I felt an emptiness, for I had always been able to talk things over with my mother, and before that, with my grandmother and Tom and Uncle Bob, gaining valuable perspectives. And Connie had always felt a great closeness and love for her parents and missed them sorely.

But we were embarked on a new course and realized that both the symbols and presences of the past were behind us. We were about to change the emphasis of our lives from the full-time practice of pediatrics to the less arduous part-time teaching, thus releasing more of my energies to planning the future of the lake property.

Accordingly I notified my associates of this decision and in May 1972 wrote to tell all my regular patients, informing them of my plans and concluding my letter:

> Although giving up my practice, I shall work part-time in specialty pediatric clinics, and in teaching or writing. There will also be more time to pursue my abiding interest in natural history by assisting in the development of a new wildlife park, and by joining with others in efforts to improve the quality of our environment and preserve wilderness areas for all our children.
>
> I have made this decision with mixed feelings and after long thought. You are my friends. We have shared and learned many things together. Your babies, your

children and, in many instances, their children have occupied a place close to the center of my life, and I shall not lose interest in nor concern for their future.

Come and see us when the park is completed.

At the same time, I offered my services to the University of Washington Medical School and was appointed a clinical professor to teach practical pediatrics to medical students at the junior and senior year levels, combining didactic sessions with demonstrations in both well- and sick-child clinics.

I closed my office the last day of December 1972 and transferred headquarters to our house on Horseshoe Lake, while the townhouse served as a Tacoma base. From this point on, our lives have been mostly lived in the midst of the park, surrounded by its wildlife and deeply involved in its joys, uncertainties, conflicts, and problems.

Land—Its Highest and Best Use

IN 1937 WHEN WE ACQUIRED Horseshoe Lake and the first portion of its environs, its highest and best use from the standpoint of the government was to get it back on the tax rolls. From the standpoint of our friends and casual observers its most obvious use was to make us "land-poor." Neither viewpoint took into consideration the pleasure and delight its possession gave us, even though we had no plans for its future development. Little did we realize how rapidly and spectacularly it would metamorphose from the drab brown chrysalis it appeared at the time to the soaring beauty of today.

At first we owned the land quite simply, watching with wonder its natural recovery. Then the land began to own us, demanding all the resources that we could muster to develop its potential—clearing fields, managing the forest for the long-term, resisting the temptation to reap premature rewards. Connie and the children were part of all this.

By the 1950s, the "large amount of unproductive land," had evolved to the state of "marginal land," with an obvious future for timber production together with an immediate though modest potential for ranching. We carried on the dual activities of raising cattle and improving the forest by judicious limbing, thinning, and selective logging until the time came in the sixties when the land and our family owned each other and a balance was struck. The new fields fattened our cattle while the growing trees of the forest, through selective logging contributed their share to the establishment of this balance. The land blossomed into adolescence, full of promise and beauty, with most of the niches available to plant colonization now filled. The meadows were well established, the forest trees were towering columns, and an understory of vegetation had begun to form.

It was toward the end of this period that a third and exciting potential of the property began slowly to enter our consciousness. We are often asked, "When and how did the idea of making the Horseshoe Lake property into a wildlife park first come to you?" Both Connie and I find it difficult to answer that question, for the concept formed gradually and from a variety of sources.

First was the realization that our daughters, all three married and with children, lived busy lives in distant places and were not in a position or even interested in ranching, land management, or lumbering. When Connie and I were gone, inheritance taxes might force a sale of the ranch.

Second, some of the trees had already reached a merchantable size. If we started seriously to harvest the timber, we would inevitably alter the face of the land and reverse its evolution toward wilderness—a transformation that increasingly awed and delighted us. How were we to reconcile this dichotomy?

Third, as the wetlands, the forests, and the meadows matured, the appearance of man's intervention lessened. As I rode my horse over the rocky ridge between the beaver ponds and swamps and through the meadowland, the grazing cattle appeared less congruous and more out of context. I found myself visualizing, instead, a herd of elk moving cautiously into the meadow from a projecting patch of timber. And when skirting the peat bog, its surface nearly choked with lily pads, duckweed, and tules, I imagined a bull moose belly deep raising his dripping head above the water, vegetation draped from one giant antler.

The ridgetop with its open areas and madrona grove, its

steep sides and lookout points, seemed suitable for sheep and mountain goats, despite the lack of screes or cliffs. Of course, native black-tailed deer already occupied the woods and brushy areas. Woodland caribou would find the combination of brush, grass, and woodland an ideal habitat. Even pronghorn might adapt to the open fields and hillsides. Shaggy, monolithic bison as well as elk would graze and keep the pastureland open. It seemed to me that this microcosm of only six hundred acres provided ideal habitat for all the species of American big game, while the gullies, corridors, and small sheltered areas furnished retreats for secretive parturition and escapes from interspecific encounters during the rutting season.

The thought of turning Horseshoe Lake Ranch into a wildlife park solely devoted to Northwest American species began to take hold. Connie and I discussed the idea together and then with our families. They, too, after a period of skepticism, were caught up in the idea and approved. We realized that such a scheme could not be carried out as a private enterprise for the same reasons that ranching or logging were not suited to our children's plans. In addition, the capital required to launch a wildlife park and provide for public facilities and participation would be too great. The need for long-term planning and continuity made it imperative that the land be turned over to some governmental body, although with strict conditions as to its use and a guarantee to us of a lifelong on-site involvement in its development and operation.

It was in 1970 that a new consideration lent urgency to the discussion. A California realtor made a tentative offer to buy the ranch for development as a community or club with stables, boating and fishing facilities, riding trails, and expensive housing. The suggested price was staggering, but this seemed to us to be the lowest and worst use for the land.

Land changes, as ours had changed over the years. We had learned that at any given time the highest and best use of the land depends on its state of development or degradation, as well as on economic factors. And, in recent years, it depends on its potential as a public resource. (Land is increasingly subject to zoning laws, laws of access, and as a last resort, to the right of eminent domain.) Our land was ripe, and we were ready to take the next step; not to develop it into an elite housing community, but to turn it over from private ownership to public ownership so that it could be preserved and serve the public at the same time. As the concern for preservation, recreational land, and conservation of natural resources increases, more land is reserved for the public sector in this way. Private donors, such as ourselves, and agencies representing the public then decide on the highest and best use of the land; whether to use it as a park, a nature preserve, or a sanctuary.

We felt, however, that we must discuss this matter once again with our children, since it had obvious long-term implications for their futures as well as ours. Each of the girls enthusiastically endorsed our plan, saying that preserving the land's wild beauty was more important to them than any profits commercial development would yield. And what better way than the creation of a wildlife park? The fields would be grazed and kept open, the forest preserved, and past husbandry and management would be rewarded, while the public would have an opportunity for a wilderness and wildlife experience. The existing barns, corrals, and other facilities were adaptable to game management; the farm machinery, vehicles, machine shop, and other equipment would be used in the operation; and we could retain our house on the lake, sharing our good fortune with others for the rest of our days.

The Wetlands

THE SMALL CABIN on Horseshoe Lake was in many ways our family hearth during the busy years of my medical practice and the children's teens. It was there, far from telephone and television, that we were able to talk quietly, read aloud (Kipling and Carrigher were particular favorites), take long hikes with our dogs, experience rural simplicity, and nourish our sense of family.

Each year, on our walks and from the windows of the cabin overlooking the lake, we could see the property becoming more beautiful. Under the benevolent and frequent drizzle so characteristic of the Northwest, the land had recovered from its former devastation and was developing the rich natural complex of biota that characterize it today. Eventually we became involved in some of these changes through our forest management practices and ranching activities, but the evolution of Horseshoe Lake, the swamps, and newly created ponds owed most to the efforts of those tireless natural engineers—the beavers.

In 1939, with our consent, the State Department of Game had introduced a pair of beaver to the lake where they quickly settled down. At first, they dug elaborate bank dens and canals, then later built a large two-story lodge against the steep shore across the lake from the cabin. As first choice for food, they cut all the willows growing along the water's edge and next attacked the cottonwoods, some of which measured more than twenty-four inches in diameter.

Next they began the construction of a dam across the small seasonal stream that drained into the lake from the peat bog. This flooded the lowlands to the east of the lake and created a new ten-acre pond, which rose to a depth of three feet, drowning the alder and ash trees that occupied this area. Marsh grasses soon invaded this new pond and a few years later cattails from the peat bog began to appear.

It was not until 1979 that the wild yellow water lilies first unfurled their succulent leaves as they reached upward from the muddy bottom. They, too, must have come from the peat bog where they covered most of the water's surface not already preempted by bulrushes and steeplebushes. In the late thirties and forties, water lilies had also grown in the shallows at both ends of the lake, but had gradually disappered where the beaver dug and ate their roots. As the watershed became increasingly protected by the emergence of the new forest and the growth of vegetation along the banks and margins, the lake and other wetlands gradually deepened and stabilized. The water level thus became too high for the reestablishment of lilies in the lake, while in contrast, their growth and spread was encouraged in the bog and ponds, which now retained some water even in the summer.

The peat bog, which is along the base of the ridge and is fed by springs in the hillside, appears to be ancient. The peat extends to a depth of at least eight feet and contains two layers of ash, one from an early eruption of Mount Saint Helens, and a deeper one, perhaps from the great eruption of Mount Mazama, when Crater Lake was formed. Until the growing forest began to provide protection for its watershed, the bog dried up in late summer, the lily pads drooped, although the sturdier bulrushes remained standing dry and brown between clumps of steeplebush and the old moss and salal-covered logs.

Now that there is some water all the year-round, these logs provide convenient elevated shortcuts and runways

Bobcat (photograph by Lance Kenyon.)

for wildlife. Before the chain link fence was built around the free-roaming area of the park in 1971, we sometimes came upon the tracks of resident coyotes, bobcats, and wandering bear or cougar that used these convenient crossings. Now only the smaller predators—raccoons, skunks, weasels, mink—and surprisingly the park deer, mountain sheep, and mountain goats continue to travel these well-worn trails, picking their way with care over the mud-daubed top of a beaver lodge that straddles the most frequently used of these log runways. Ducks rest on them in rows to sun themselves and preen. They provide nesting sites for other waterfowl, especially the Canada geese, and muskrats drape rushes and pond grasses over their semisubmerged ends.

Beaver are still the most conspicuous inhabitants and modifiers of our wetlands, for they never let up their vigilance, repairing lodges and leaks in dams, cutting trees and excavating canals. There remain, however, factors that they are not able to control in the areas where dry land and water intermingle so intimately, becoming transmuted as the water table rises and falls with the seasons. Then, despite their best

California bighorn rams. (Photograph by Gary Oberbillig.)

engineering efforts, the peat bog and pond become too shallow for underwater swimming. The beaver are forced to concentrate on the original two-story lodge on Horseshoe Lake, which has been in constant use since its construction by the original pair of beaver.

How three colonies of beaver accommodate to one another now that the area is tightly fenced and outward migration is blocked is not clear, since normally each pond is occupied by a single family. I am sure that before the fence was built in 1971, preventing the dispersal of the two-year-olds, there was no problem with overcrowding. But now that their canals to the outside and their old trails toward the Ohop Valley are blocked, and since they are all related, they must have gained some tolerance for greater togetherness. When the lake is very low we see their complex of bank burrows extending thirty or more feet toward the Oregon ash and alder stands that rim the south shore of the lake and terminate in well-concealed air vents beneath a bushy hazelnut or pile of brush. These dens probably offer relief from the crowded conditions of the low water periods.

At the wettest time of the year, the lake, the swamps, and bogs now constitute 10 percent of the surface area of the entire property, shrinking to about 5 percent before the heavy autumn rains replenish the thirsting shallower depressions. All these changes in the wetlands have enhanced the habitat for great blue herons and tree-nesting ducks of all four species. A bittern found a home in the rushes, and occasionally I see a green heron in the same area. Of course, legions of water insects and larvae, frogs and salamanders, garter snakes, and turtles have all benefited. In winter the wetlands are silent resting places, but in spring and summer they teem with life.

Many scientific and popular books have been written about life within these shallow fecund bodies of water. They hold a poetic and primitive fascination for most people, particularly small boys and girls, but also for older naturalists. Armed with collecting bottles, sieves, nets, a magnifying glass, and no aversion toward slime, mud, or pungent and unfashionable odors, one can begin to glimpse the wonder and exuberance of these habitats. The surfaces of floating leaves and blades of grass and sedges form platforms and perches for insects, frogs, and other amphibians. Beneath their surfaces and on their stems the larvae of dragonflies, hydra, and other fierce predators lie in wait or stalk their prey as intently as a cougar or weasel. Even on its surface the water tension supports whirligig beetles, water striders, and scorpions. Floating duckweed afford food for waterfowl, righting themselves if overturned by wading moose or swimming duck because of the water-repellent nature of the upper surface of their leaf. Myriads of microscopic forms of plants and animals occupy the mud and circulate through the temperature currents of these shallow waters. The study of limnology is one of the most fascinating of all of the natural sciences, but for us here it is enough to appreciate the significance of these wetlands for they form the wide base of a great food pyramid.

Two other bodies of water deserve special attention because of their importance and because their origins and characters are different from the swamps, bog, and beaver pond. The first of these is a pothole, or "Pothole Lake" as it is now called. It is a shallow weed- and grass-filled pond of about three acres, surrounded by large cottonwood trees and Oregon ash. At its edges stranded or floating moss- and grass-covered logs form a crisscross, which is barely negotiable by curious human explorers. Many of these logs are

White-tailed deer buck in velvet.
(Photograph by W. H. Shuman III.)

cedar and still sound even after fifty years of alternate immersion and grounding. Geologists speculate that this small lake was formed during the last glacial age by a large chunk of ice which, after creating a depression, slowly melted. Over the centuries the clear sterile pond gradually became colonized by vegetation and abundant animal life. Silt, leaves, and decomposing plants have continued to accumulate, and as the water becomes shallower, it is choked increasingly by lilies and other water plants. Pothole is a lake dying as all lakes die eventually, and it will soon transmogrify to a swamp, then a wet meadow, and some centuries later an alder flat.

In the meantime it provides a favorite courting and nesting area for tree-nesting ducks: buffleheads, goldeneyes, wood ducks, and hooded mergansers. These birds find protection amidst its floating logs and water plants and nesting holes in the rotting snags and man-made nest boxes in the nearby forest. The warm shallow waters of this weedy pond also provide ideal habitat for newts, salamanders, several species of frogs, water insects, and small leeches; good hunting for coon, mink, skunks, and heron; and good habitat for muskrat, voles, and shrews.

By far the most important body of water is Horseshoe Lake. The lake is the reason we acquired the property in the first place. It is the focus around which the later ranch was organized, and it is the heart and most attractive area of interest for the visitors. The lake has also been the center of interest for our family, for it is the foreground of the view from the cabin and later the main house. Not thirty feet from our door, the fern- and salal-covered bank meets the water's edge. The curve of the lake enfolds the house, and you can glimpse its rippling water between the trunks of the tall fir trees that have replaced the brushy growth that was there when first we saw it.

In 1949 there was a severe earthquake in the lower Puget Sound area which apparently broke the clay seal in the bottom of the lake. All the water emptied out dramatically almost overnight, leaving in its place a dark, evil-smelling mud flat. More smallmouth bass, crappie, and large catfish than we ever dreamed could coexist in such a shallow lake were stranded and died. The word got out somehow that there was a banquet laid out at Horseshoe Lake. Within a few days, raccoons, seagulls from the Sound, and all manner of other scavengers moved in and cleaned up the mess, but the dead, ugly brown depression remained unchanged all summer. It was a shock to see our lake vanish and with it much of the basis for our future plans.

We need not have worried, however, for with the fall rains, the lake started to fill up again and by spring it appeared just as it had before the quake, except that it seemed devoid of life. Again we were misled, for that very next spring, I noticed a sinuous dark shadow about two feet in length and half as wide moving slowly along the shore and recognized it from previous years as a school of black, large-headed, bewhiskered baby catfish. This time they were orphans for there were no adult catfish left to guard and shepherd them as is the natural pattern. These inch-long babies were obviously hatched from eggs surviving deep in the mud of the empty lake.

Whether the trout, which the state insisted that we plant, compete with the catfish I do not know, but in the more than thirty years since the baby catfish appeared, they have grown only to about ten inches, whereas some of their forebears stranded in the mud of the dying lake were twenty-four inches long. The trout, on the other hand, grow mightily and become pink and fat on the fairy shrimp and abundant insect life. We had fine fishing for many years, replanting from time to time because there seems to have been little if any spawning in the lake. We have made several new plants of fish in the last ten years, however, there has been no fishing since the property became a park. Yet quite a number of large hook-jawed old rainbows still cruise the deeper waters of the lake and take their toll of their smaller fellows and of the catfish and bullfrog tadpoles.

The lake water is deeper and clearer than in the early

Wood ducks. (Photograph by Gary Oberbillig.)

days. Now the peat bog and the beaver pond act as reservoirs and a stream flows from them down through the meadow and into the lake throughout the winter months until June. We also drilled a well near our house in 1965 and can pump additional water into the lake in large quantities during dry seasons or provide open water for the waterfowl when the lake freezes. In addition to the rainbow trout, catfish, and bullfrogs in the lake, western painted turtles sun themselves on the logs, but they too have been introduced during our time here.

I never tire of watching the lake. Ripples scamper across its surface when there is no apparent wind. Logs float gently back in the evening and forth in the morning with the seiche. A trout rises in swirls of liquid light. Catfish mug along the surface when the air is still. Ducks fly in and out. Geese talk their way down over the trees to land with set wings and braking feet, then congratulate one another loudly on the success of their adventure. Whistling and trumpeter swans glide with one foot tucked along their side, scarcely rippling the surface. The lake is never the same from day to day and hour to hour, for its surface mirrors the mood of the beholder and yet excites and stirs his imagination if too prone to indolent reflection.

The Forests

MOTIVATED BY THE SPEED with which our cutover and burned-over land was returning to forest, we began by 1956 to plan seriously for the development of our six hundred acres surrounding Horseshoe Lake. During our first twenty years of stewardship we had done little with our land except observe and enjoy its natural regeneration from the cataclysms of the twenties. Now the time had arrived for us to take control and responsibility for further development. The forest needed management, and the brushy areas were ready for clearing and seeding into pastureland for our embryonic ranching operation. Already the moister portions of the woodland had grown up into nearly pure stands of western red alder, eight to sixteen inches in diameter, well spaced, and free of lower branches. Had the land been less moist, these would have been overtopped by fir trees, but the latter had made little comeback, and the hemlock and western red cedar that had succumbed to the loggers and the fire were slow to reestablish themselves. It seems unlikely that in these alder stands the original forest will ever be restored.

Not too many decades ago alder was considered by most foresters as a weed tree—now we know better. Dry alder burns brightly in the fireplace with little sparking and only a small residual of fine white ash, while when still green it is an ideal fuel for smoking and barbecuing meat—especially fish. Much of our alder, cut while still undried into dimensional lumber, is shipped to furniture factories in California and elsewhere and manufactured into light and inexpensive furniture, sometimes labeled "maple." Others who have carved alder say that it is easy to work and has a fine grain and texture. Perhaps the least appreciated, yet most important quality of western red alder is its ability to fix nitrogen in the soil. Like a legume, it has clusters of tiny orange bacterial nodules on its rootlets that performs this vital function. After logging or fire, small alders appear in great numbers almost as if by magic, restoring nitrogen to the depleted soil and creating an ideal seedbed. If the ground is not too moist, Douglas fir seedlings will overtop the alders between thirty and fifty years later, cutting off their source of light and killing them. The dead and fallen alders then rot rapidly, restoring humus to the forest floor and sometimes contributing one more bounty—the shell-like, pure white fungi called oyster mushrooms that thrive almost exclusively on the decomposing alder and are so delicious lightly sautéed in butter. Obviously we no longer look with disrespect on the alder in the Puget Sound country.

The Douglas fir, on the other hand, did not grow nearly as uniformly as the alder. In some areas thickets of tall firs, only two to four inches in diameter, were so crowded that none could gain dominance and hence useful size, while in others, trees of the same age but better spaced, had reached a girth of twelve to sixteen inches. Even in these more favorable stands, however, there was the usual struggle and competition for light at the crowns of the trees.

Several government programs for cost-sharing on limbing and thinning projects were in operation in the fifties, subsidizing 70 percent of the expenses in the interest of improving timber production on private lands. We qualified for such a program, and the State Department of Forestry made studies of our fir stands. Eventually we arrived at a long-term plan for forest management. Whenever we could afford it we carried out that plan during the succeeding years.

Ignoring the dense thickets at first, we concentrated on the better areas. Neither limbing nor thinning are pleasant jobs. First one must assess a stand of Douglas fir that consists of trees all of the same age, even though their size may vary markedly according to how much light reaches the crown of each tree. Consequently, the most dominant trees stand out to the practiced eye. They are usually spaced from fifteen to twenty feet apart in a typical twenty-five- to thirty-year-old forest. Between them, and perhaps twice as numerous, are well-formed but smaller trees spaced eight to fifteen feet apart that are considered reserve trees. These are capable of replacing a dominant tree should it be destroyed by storm or other damage. As the forest grows, the reserve trees are logged and sold when they reach merchantable size, eventually leaving only trees of fine quality, fairly uniform size, and evenly spaced in each managed stand of Douglas fir. Of course, in addition to the dominant and reserve trees, there are many smaller spindly and misshapen ones that will fail in the struggle for light, but meanwhile in the young forest they compete for nourishment and slow the growth of the more viable trees.

In the first phase of management, we eliminated these inferior trees. Removing them, however, proved too expensive in those days. As there was little demand, even for the slender straight specimens as hop or corral poles, they were simply cut and left to rot. Meanwhile, we limbed the dominant and reserve trees to a height of at least eighteen feet. This was done most rapidly by knocking off the dead lower branches with an ax or club and cutting off the higher, still living branches with a pruning saw. Sawing limbs overhead while dry fir needles and bark shower down into eyes, sleeves, and shirt collars is hard work, but if done properly, the wood and bark of the pruned trees grow over the knot and in time provide a beautiful sixteen-foot log of clear, close-grained lumber. We managed thirty or forty acres in this manner while limbing dominant trees over a much wider area without doing any thinning. In 1969 we contracted with a logger to take out the now merchantable reserve trees and some others from the areas that had been less intensively managed. He set up an efficient small mill near Pothole Lake cutting cants for about four months. Revenue from the sale of this timber, as well as from the fat steer and heifers nearly eliminated the need for infusions of capital, while the prospects for continuing profits from future timber sales seemed bright.

It has been fascinating to observe the appearance of new species of vegetation as the forests have matured, for this property, now called Horseshoe Lake Ranch, has a definite history to mark the stages of its recent development. The story begins with its logging in 1920, when most of the fir and cedar were removed, and continues with the Great Fire of 1924 that completed the decimation of the forest. Only the occasional fir tree or cedar protected by surrounding swamp or nestled in a ravine was spared by the advancing flames. From the seeds of these scattered trees the forest was restored with the help only of the wind, the Douglas squirrels, the Townsend chipmunks, and a few species of birds. Thus it was possible to chronicle how this microcosm of Western Washington containing most of the land forms typical of the area has, like the Phoenix, risen from the ashes and sequentially, over a period of slightly more than fifty years, become the beautiful, reforested wildlife park that exists today. The emergence of the Douglas fir forest to a stage of late adolescence has been hastened by forest management. Already feathery hemlock are beginning to grow in the shade of the firs, foreshadowing their emergence as the dominant species in the climax forest many years from now if nature is allowed to have its way. The alder forest is now nearly mature, and the wetlands are established. The underbrush, consisting mostly of salal, Oregon grape, and huckleberry, has been reduced by the browsing of cattle and now elk, deer, and moose. In comparison, the aquatic plant life has burgeoned with the appearance of new species and the spread of the lilies and rushes.

Until the late fifties there was virtually no sword fern growing on the property, but now the forest floor is carpeted with it wherever there is sufficient light. In 1978 the first licorice fern (polypody) began to grow on the moss-covererd trunks and branches of the now mature big-leaf maples.

In 1980 I saw the first devil's club by the edge of the bog, its huge maplelike leaves and hairy, wickedly spiny stems rearing upward to catch the filtered light beneath the cedar trees. Later in the spring, bunches of white flowers, then red berries give an exotic appearance to this armed and dangerous plant, which only the elks seem to be able to browse with impunity. Other new colonizers included maidenhair fern, which I first noted in 1970 along the north fence line and on the wet hillside above the Ohop Valley. Scattered trillium began to raise their lovely, three-petaled white and purplish flowers above the forest floor about the same year.

As this colonization by new species continues, other trees and plants have passed their prime and are decreasing

Elk bulls vying for herd leadership.
(Photograph by W. H. Shuman III.)

Moose. (Photograph by W. H. Shuman III.)

in numbers. The black willows growing on the flats atop the ridge are dying and most are now too tall for even the moose to reach their few remaining living branches. Many of the hazelnut bushes have also grown too high to afford good browse, whereas cascara trees lack vigor and are being crowded out. The madronas, however, are now forming a magnificent grove on the ridgetop, and the moose eat their bark. The Oregon ash has been nearly eliminated by the beaver, and the wild dogwood is under attack by a fungal disease anthracnose and many will succumb.

Only an occasional spindly vine maple, fern, red huckleberry, sparse salal or wispy Spiraea survives the pervasive shade and acid soil beneath the pure fir stands. Such stands are little used by wildlife except as shelter from heavy rains and snow, although Douglas squirrels, and to a lesser extent Townsend chipmunks, find them favorite habitat. When the fir cones are abundant—green and sticky—a single squirrel may cut a hundred or more cones in a day, storing some for winter use. The industrious human who retrieves these cones can earn a tidy sum for a sack of good quality ones.

Finer cones command a yet higher price for seedling production.

Genetic manipulation and selection in the research facilities of our giant lumber companies are now producing trees with incredible potential for growth and rapid maturation. The coarseness of their grain and roughness of their texture, however, can never duplicate the uniformly beautiful wood of old growth timber. At one time a poet assured us that tree making is a divine monopoly. Now there are many humans in the business.

Undergrowth is vigorous and varied in the mixed forest areas. Among the scattered firs, cedars, hemlock, alder, and occasional spruce and dogwood, the large Oregon or bigleaf maple trees thrive. They are aptly named, for some of the leaves on the young shoots attain a twelve-inch span. Other deciduous species are interspersed throughout the mixed stands. Vine maples form a delicate tracery above the forest floor, green in summer but a scarlet splash in the fall woods. Both red and blue elderberries grow where the soil is moist. Cascara, Oregon ash, and black willow shade the canes of blackberries, gooseberries, salmonberries, and thimbleberries, adding a variety of forms and tints to the understory. In the wetter areas, large cottonwoods are plentiful, perfuming the spring air with the scent of balsam of Peru from their sticky buds and later, in the fall, adding the clear yellow of their fading foliage to the pageantry that accompanies the death of leaves.

Horseshoe Lake Ranch

HORSESHOE LAKE RANCH was officially born in April 1958 with the purchase of two mature registered black Aberdeen Angus cows, each with a heifer calf alongside. Although the lake property had been peripherally fenced since 1942, there were no cleared fields but only scrubland and small open meadows to provide grazing. We would have to clear and till sufficient land to provide good pasture during the spring, summer, and fall months. Other decisions were not so clear. Should we start a cow-calf or feeder operation—buying calves in the spring and selling them in the fall? Should we raise or buy winter feed? Would irrigation be required?

I had been around cattle and spent time on various ranches, but my experience was limited. It is true that we had kept a few dairy cows at the lake, but raising beef was an altogether different matter. The advice of our county agent, who knew local conditions well, proved invaluable. I also did a lot of exploratory reading of books and government publications on cattle management. I did not realize at this time that the lessons we were to learn—sometimes painfully—about the management of domestic cattle would prove in the years ahead to be highly applicable to the care and control of wild ungulates.

The soil in our area, which is largely dominated by undulating and somewhat poorly drained glaciated uplands, is not suitable for intensive agricultural use. Where the landscape is rolling, water tends to stand in ponds during the winter months, and small lakes such as Horseshoe and Pothole dot the landscape. Elevations vary at Horseshoe Lake Ranch from four hundred feet to a thousand. The annual rainfall is between thirty and forty-five inches and the mean temperature is about forty-five degrees Fahrenheit. The average frost-free season is one hundred and eighty days with good grazing available for the two hundred and twenty days or so between mid-April or the beginning of May and mid-November or a little later. The native vegetation is predominantly Douglas fir, western hemlock, western red cedar, and alder. This association of soils is suited to raising hay, to pasturing, wildlife habitat, recreation, and (perish the thought) urban development.

Raising and making hay on the west side of the Cascade Range is, however, risky. Rains are common through June and even later, so that often it is not possible to mow until the grass has already gone to seed. Even then unexpected rain may soak the hay before it is cured, baled, and in the barn.

We finally decided to develop our ranch as a cow-calf operation, depending on pastureland for spring, summer, and fall grazing while winter feeding with good alfalfa hay from east of the Cascades. As some of the best woodlands in Western Washington are in this area, we also decided that forest management should be practiced in conjunction with the cattle operation. With that in mind, we would clear only the flattest land for pasture where the fir trees were small and sparse but where brush and alder were thriving. Accordingly, we chose two main areas, each about twenty-five acres, to establish our permanent pastures. We began by clearing the area between the lake and the bluff overlooking Ohop Valley and extending behind the ridge until it and the bluff began to merge. This was probably the least fertile soil on the property.

We bought a TD-18 bulldozer for twenty-five hundred dollars in 1956 from the American Salvage Company. It was an ancient navy surplus machine with a dirt blade raised by cable and lowered by gravity so that one could not dig with it as successfully as with a hydraulic blade. Charlie McKasson, who with his wife Ila and children had replaced the Calverts as caretakers on the property and now helped out with various projects, built a clearing blade with teeth out of used armor plate and railroad rails to dig and comb out the roots and debris left from the logging of 1920. Since most of the alders we cleared were at least nine inches in diameter at the small end of a sixteen-foot log, they commanded a good price at both the furniture factories and the pulp mill. This helped pay for the fuel and repairs on the bulldozer, which without Charlie's skills and imagination and our welding equipment would never have held together long enough to complete the clearing job.

As the alders were tipped over, limbed, and the roots cut off, and the fir stumps rooted out or split and shaken loose with dynamite, we started huge fires that burned continuously as we piled more and more debris on them. Weekends when I was not on call I worked along with Charlie and the machine, picking up limbs and roots and feeding fires. That done, we gathered cobblestones endlessly and gradually leveled the field, covering the gravel with topsoil where it had become exposed as the stumps were rooted out. When this area was finally cleared, fenced, and seeded, we started on the second twenty-five acres which lay mostly in the alder flats east of the lake. Here, where the land was more rolling, we left the old stumps, thus not disturbing the topsoil as much. A year later we were ready to fence and plant this second pasture.

In March 1959 we fertilized the land for the first time and seeded it to orchard grass and New Zealand white subterranean clover. Our goal was to establish a pasture of about 60 percent grass to 40 percent legume. Before seeding we fenced off the cleared land from our growing herd of cattle, dividing each of the twenty-five-acre pastures in two, thus forming four fields of roughly equal size, each opening with a gate on a lane that led all the way up to the corrals and barns near the McKassons' house. This system made pasture rotation and cattle handling simple and efficient, for one had only to open the gate to the field into which the cattle were to be moved, and then, on horseback or on foot, herd them into the lane, closing the gate after them. They then found their own way into the new pasture. The lane remained open and gave access to watering troughs, salt, and bone meal feeders.

The corrals were designed with small but adequate holding pens. Cutting gates channeled the animals into the chute, then into a cattle squeeze. From the squeeze, other cutting gates directed the calves and cows into a second holding pen and back out into the lane, or through a scale for weighing, or up a loading chute, or into the feedlot and sheds. We hired little outside help, and Charlie did the work of two or three men by himself, while if the job required three or four men, I helped him on weekends and vacations. It took about four years to finish the barns, concrete paved feedlots, and other structures. We were proud of the efficiency and ingenuity of our facilities as good corrals are vital to efficient stock management, saving man and beast from unnecessary stress and danger.

During winter and early spring we fed the cattle good second cutting leafy alfalfa hay from east of the Cascade mountains and a grain ration containing about 16 percent protein. At this time of year they were excluded from the pasture and roamed the woodlands and brushy hillsides, browsing the salal, sword fern, and what huckleberry they could reach on the tops of the stumps. This browsing altered the natural flora of the forest floor by nearly eliminating the salal, huckleberry, and sword fern. The huckleberry has made a slow comeback, and sword fern now carpets the forest floor, but the salal will probably not recover as it is cropped by the deer and elk as soon as it begins to appear on and around the fir stumps.

We always fed from our flatbed cab-over jeep, dropping

off the unbroken flakes of hay in widely separated piles on the well-drained ground around the pasturelands. We fed at the same time of day, summoning the stock (if they were not already waiting) with the honk of the horn, although they soon learned the sound of the engine and would begin to gather before we reached the feeding area. Everywhere, except in the newly created pastures, alder spread. Tiny seedlings in dense clumps preempted open ground. For some reason, which I do not fully understand, there has been no encroachment on the fields, even though they are bounded in places by alder trees.

Each year, after the initial fertilization of the fields, our routine has been to fertilize as required and simultaneously to pasture harrow the fields in mid-March, weather permitting. If the clover was abundant, we used no nitrogen, but if the grass was crowding out the clover and reducing its proportion to below 30 percent, we manipulated our fertilizer mixture, increasing the potash and sulfate. This served to regulate the composition of the pastures. From time to time we checked our hunches by soil testing.

Along fence lines and openings not yet overgrown with alder, I seeded grass and clover with a small hand-turned broadcasting seeder. This was a satisfying occupation as I paced with measured tread, turning the handle and watching the thin, gray-brown grass seeds spray out in a wide arc and hearing the tiny dry susurrations as they settled to the ground. The fewer and more precious red-gold pellets of clover seed fell with an even fainter patter where leaves lay or the ground was hard.

In the swamps and bogs I seeded reed canary grass from horseback. When my new horse, Cherokee, and I became accustomed to the flying seed drifting into his and my eyes and ears, under his saddle blanket, and down my neck and boots, we found that we could reach most of the wetlands in this relatively easy manner. It took several years for the reed grass to take hold, and even now it is scanty, but its delicate, tufted heads and tall, coarse stems and leaves are ornamental and provide fairly good feed.

How many of you, I wonder, have ever seen or held clover seed? It is not only beautiful in the tiny perfection of its roundness, but somehow it epitomizes the wonder of germination, renewal, and enrichment of the land. Clover seeds are hard and separate, yet run through the fingers like mercury. They seem to be of the temperature of one's own flesh and nestle in the palm weightlessly so that one must look to see if they are still present. For a small sum, you can buy a few ounces of clover seed. Even if you have no place to sow it, keep it as a treasure in a clear glass vial, and when the world around you seems empty and meaningless, you can take the golden seeds, tip the glass, and watch them flow. Pour some in your hand—you may have found a talisman.

Cattle Ranching

RANCH LIFE–WITH ITS romantic association of cowboys and horses, shoot-outs and Indians—has come to represent some sort of uniquely American ideal, a life-style endlessly re-created in fable and fiction. Although the realities of ranching are far more prosaic, my fifteen years in the cattle business at Horseshoe Lake Ranch, my many contacts over the years at various times and places with cattle raising, and my continuing sporadic involvement with Wyoming ranch life, as lived by my nephew Rob Hellyer and his neighbors, have been a source of deep satisfaction and useful expertise.

Cattle ranching is not all romance, although there are occasions throughout the year when work is not pressing, fences are in good repair, the grass is thick and green, and the flies not too troublesome. When cows graze contentedly or lie peacefully chewing their cuds while one of their number babysits the calves, you can overlook this peaceful scene from the saddle with a feeling of vast contentment. A cattleman's life includes other high points: when all the steers and old cows have been sold for a good price; when your replacement heifers are well grown and uniform; when your cows have calved out except for a very few late ones; and when spring grass is well up in the pastures and rangeland before all the winter hay is gone. At such times the rancher knows he is in the right occupation and would not trade his life-style for any other.

Ranch life has its own calendar of events that does not correspond with the national celebrations such as July 4 or days of reckoning such as April 15. These events are, however, controlled by cycles based immutably on daily, monthly, and annual biological rhythms that cannot be disregarded. The circadian rhythms of the ungulates, whether domestic or wild, center largely around periods of rapid and steady eating during the early and late hours of the day and then the quiet times of fermenting, ruminating, and digesting this forage during the middle hours of the day and night. Monthly rhythms are concerned with short periods of receptivity by the female to the male and reproduction, while seasonal rhythms controlled largely by the length of daylight affect the overall hormonal activities of most living things. Domestication has modified natural reproductive cycles since man provides yearlong food, protection, and shelter for his animal dependents; thus breeding time in domestic animals is not restricted to certain seasons. The rancher can improve his own patterns by permitting breeding only in the late spring and early summer so as to obtain calves the following year when abundant grass of good quality—no longer washy and not yet too fibrous—provides ample milk for the rapidly growing calf. Once the rhythm has been established, the pattern cannot be changed by decree as in the case of Thanksgiving.

The rancher's calendar is much the same whether he is a small operator with a closely controlled purebred herd like ours at Horseshoe Lake Ranch or a cattleman with a thousand head grazing over miles of open range. It consists of a few major events and many minor chores, all of which determine the success or failure of the entire operation. The first and most important of these events is spring calving, which is timed to occur in most areas of the Northwest in February and March with a few late calves inevitably dropping in April. The more closely synchronized the calving, the more uniform the calves, and the less drawn-out this anxious period.

In times past, calving occurred on the open range, and the rancher had little control over the outcome. Some ranchers still accept this method, but most cattlemen nowadays bring in their cows close to the ranch where calving can be observed and supervised when necessary. Our cows at Horseshoe Lake Ranch calved in the woods or clearings unless we were concerned about a particular young heifer or older cow with a history of previous difficulties, and these we brought into the corrals.

When at last calving is over and the calves are nursing, growing, playing, and have joined the herd, the rancher decides whether to sell the few late calving and open cows or hold them over. But relief and satisfaction are the prevailing moods at this time, and the ranch family can take a few days off for recreation and self-congratulation.

The next important event in the cattleman's calendar is the roundup followed by branding and all the other chores that go with it. "Gathering" consists of rounding up all the cows, calves, and steers and driving them to the corrals. Most ranchers have already brought the cows that are due to calve in close to the main buildings where they are being fed daily. It is not difficult to round them up, although usually a cow or two disappear into some gully or brush patch and have to be chased out. In our case at Horseshoe Lake Ranch, once in the lane that leads from the field and range area to the corrals, the cattle could be trailed by one man on horseback or even on foot, but there is really no substitute for a horse when gathering cattle. On big ranches where cows are widely scattered and the terrain rough, gathering is a project for all the neighbors who help one another with these spring chores.

When each rancher's time for branding comes, his neighbors arrive at about eight in the morning. At the home ranch, all must be in readiness because the neighbors are prepared to put in a day's work so arduous that it is beyond price. They provide their own horses, their special skills—to rope or ride or wrestle calves—and soon the corrals are full of bawling cows and milling offspring. The horses used for gathering the cattle are tied to corral fences, and the hard work begins. This is the time when husky, enthusiastic high school football players are at a premium. Otherwise the oldsters must bust the calves themselves, get kicked, bruised, and tromped. But first, the cows must be cut out from their calves. In a good corral (not too big and with rounded corners) this is not too difficult if the man at the cutting gate is experienced and can think like a mother cow. This accomplished, the scene moves to the area where the branding will be done. Nowadays branding irons are usually heated by the roaring jet flame of a propane torch instead of by a wood fire.

Some brands are simple and can be applied with a single iron, while others require the addition of bars or other figures. Brands are proud possessions—unique and registered in each state. A good brand may sell for thousands of dollars if the owner has no further use for it. Our brand is a horseshoe with an H inside and is still registered, even though I will probably never again stamp it on a critter. Branding, by common consent, is a job for the owner of the cattle. If he places the brand upside down by mistake or smears the job, it will reflect on him and not on a neighbor, and you can make a joke of it.

And so it goes until lunchtime. Iced tea, lemonade, and pop are the drinks provided during the hot, dusty work in the corrals. This is not a time for beer—that can come later when the work is done. Lunch, however, is one of the ways the rancher and his wife can say thank you for the morning's labors. Boots come off and everyone washes up. The meal is always superb, with meats, different kinds of salads, pies, and cakes. Then, with sighs and groans of satiety, it is back to the corrals or on to the next ranch to repeat the same procedures. Often, after the calves have been branded, there is work to be done on the cows. Here the chute and squeeze come in. A host of small cleanup jobs are performed: spraying or applying fly medication or replacing ear tags that have been lost. Then, the reunited herd is returned to its feeding area with far more cooperation on the part of the cows than when they were gathered earlier in the day.

When working our larger calves in the corrals we put them through the chutes and into the squeeze. There we found we were able to immunize, brand, ear tag, and tattoo the ears of the heifers with identification numbers for later registration, all without the need to bust them. The small calves we manhandled in the usual way. Since Angus cattle are polled, which means that they are genetically hornless, the traumatic business of dehorning did not arise. Instead of castrating, we used the elastrator with two rubber bands for insurance—rather like wearing both a belt and suspenders—and were satisfied with this method as it seemed far less stressful than the knife. The cows were then checked, sprayed for parasites, and turned out with their calves to the newly greened and lush pastures after a full feed of hay to reduce the likelihood of bloating. The bulls went out with them, and the young heifers were kept in a pasture near the barns, well away from the bulls.

In the fall, the cows and calves were again brought to the corrals, checked and medicated when necessary, and the calves kept in the corrals to be weaned while the cows were returned to the pastures. Weaning was a noisy time, but the farther apart you can keep the cows and calves, the less distressing it is. Little sleep was to be had during those few nights as the calves and cows bawled continuously, but soon the cows dried up, and the calves went on completely solid feed. The bulls, which were brought in after three months with the cows, were put in a pen. The cows and calves and heifers were reunited while the steers were brought into the feedlot to be fattened through the winter. All the rest of the herd was turned out onto the range where they would remain until time for calving and the spring roundup came again.

During the years at Horseshoe Lake ranch and elsewhere, I also learned about ruminents in general—goats and sheep as well as cattle. The similarities between species of ruminents, whether wild or domesticated, are far greater than their differences. Both their reactions and behavior are conditioned by the highly evolved and particularly complex digestive system they share, setting them aside from equines and other ungulates.

The differences are those of adaptation. The domestic ruminant is selected for docility and reproductive flexibility, rapid maturation, and production of large quantities of milk and meat for man's use. The wild ungulate must depend on superior adaptation to differing environments in order to survive the rigors of harsh climates, sparse food supplies at certain seasons, predators, and hence retain an elegance and grace, an alertness resulting from a nervous system highly attuned to external dangers. Its behavior remains dependent on the cycles of the seasons, rutting, antler growth in the case of cervids, and varying patterns of mothering and bonding. For these and many other reasons the wild ungulates are far more fascinating than their domestic counterparts. Yet it must be remembered that goats in particular become feral very readily and can survive in nearly all environments. Furthermore, many an old cow or bull, missed during a fall gathering, survives the bitterest of winters through retention of instinctive wisdom of wild progenitors.

Ranching is not all concerned with cows, although they are the heart of the matter. Ranching involves fences and barbed wire that coils and cuts and has a twisted nature that makes it difficult to manipulate. If like me, you have difficulty working with gloved hands, barbed wire is even fiercer. Often I have looked at my scratched and wounded hands gently palpitating the soft pink abdomen of an infant and had to explain their condition to a mother looking on with curious fascination. Ranching is also building things, fixing things, inventing things. If one is fortunate, as in my case, to have a foreman and friend such as Charlie McKasson, who can do all of the above and more, one soon accumulates a store of useful junk—old car parts, used brick, bathtubs, pipes and fittings, water tanks, sheet metal, angle iron, broken machinery, old wheels, and tires. You can hide some of this wonderful assortment out of sight, but much of it just lies about where it may be needed. Working ranches are not conspicuous for their white painted fences, manicured yards, and impressive architecture. Barns and outbuildings and corrals are more functional than things of beauty, although it is just as easy to make a new barn or shop pleasing to the eye if one believes as I do that form should follow function. I am suspicious of a ranch or farm where everything is too neat, for I feel sure that treasures that will be needed someday have been thrown away, and that the boss spends a lot of time looking for something he had once, needs now, and cannot find. A piece of junk may become a treasure in seconds at the proper time.

Ranching is hard work. Luck and the elements and the banker are the cattleman's partners for better or for worse. Ranching is also a splendid way of life for one who understands and accepts these conditions yet wants to be an independent man. Yearly he views the full cycle from breeding to calving to weaning, then starting all over again. Next to medicine, ranching and natural history would be my choices as careers. I have been fortunate to have found them all compatible.

The Making of a Park

THE REALIZATION OF OUR dream of a publicly owned wildlife preserve and park at Horseshoe Lake proved far more difficult than we had ever imagined. Many times in the next few years our faith in the Biblical affirmation that it is more blessed to give than to receive was severely tested. Nonetheless, we began with high optimism, and, as it turns out, naïveté, to win acceptance of our land and of our concept for its highest and best use.

We first approached the Pierce County commissioners, thinking that our chances of maintaining local and significant input would be much greater than if the wildlife park were owned by the state. The commissioners appeared quite interested in our proposal and requested that an economic feasibility study be undertaken. We commissioned such a study and on May 10, 1971, the highly favorable findings were presented.

There followed an interminable series of conferences, postponed meetings, missed appointments, and in one instance, a last minute cancellation of a conference with the economic researchers who had come all the way from Los Angeles specifically for that purpose. I felt at times as though I were humbly trying to peddle an overripe fish. Ultimately, we and the commissioners both concluded that the county simply did not have the needed commitment or means to take on a project of this magnitude.

I therefore withdrew my offer to the county, paid for the economic survey, and on May 12 submitted the identical proposal to the Tacoma Metropolitan Park commissioners. I included the conditions of the gift as outlined in a supplement to the deed and a proposed operating agreement in which the Tacoma Zoological Society would play an active role. Before the Metropolitan Park Board had come to a decision, Connie and I decided to gamble on a favorable outcome. We began to build an eight-foot-high chain link fence around the four hundred and forty acres of property that seemed most suitable for a free-roaming area for Northwest American ungulates. In order to pay for this three and three-quarters miles of fence, we held a well-advertised auction at the ranch and sold all our Angus cattle in one day at a very favorable price. Our bridges were now burned, but the money from the sale paid for the fencing.

As anticipated, Tacoma's Metropolitan Park Board had unanimously voted, on August 23, 1971, to accept our gift (including the land, house, and farm equipment). The fencing was completed by September 14, 1971. The grass in the meadows was tall, the corrals had been adapted for game animals, hay was in the barn, ancient though adequate equipment was on hand, and only the introduction of the animals remained before the wildlife park became a reality.

On October 13, 1971, six bison from the National Bison Range in Montana were contributed by the federal authorities, and they adapted rapidly to the unaccustomed lushness of the pastures. Two white-tailed deer came next, then in October, our first moose, named Chocolate, arrived amid a shower of both local and national publicity. In February 1972 the first three Roosevelt elk were transferred from the Point Defiance Zoo in Tacoma. In addition, eight native black-tailed deer were fortuitously enclosed while fencing was in progress. Already five of the contemplated eight species of ungulates were represented. Connie and I, living in their midst, began our studies, recording observations on animal behavior and interactions in a journal, which we

Bison cow and calf. (Photograph by Karen Denman.)

continue to keep to this day.

The key conditions that were included in the document transferring the property to the park district were as follows:

> The conveyed property is to be used for the following purposes and for no other purposes whatsoever:
>
> To exhibit, propagate, and preserve predominantly native Northwest and Alaska species of wildlife in their natural habitat and in ecological combinations.
>
> To provide a park surrounding the wildlife enclosures, which park is to remain a wild area for the education, recreation, and enjoyment of the people of Pierce County.
>
> In recognition of his special knowledge of the property and interest and expertise in the field, Dr. David T. Hellyer will be given a planning and supervising position, and a nominal salary (one dollar per year) to assist in the development of the park.
>
> Duration. The terms, covenants, conditions, re-

Roosevelt elk. (Photograph by Gary Oberbillig.)

> strictions, easements, benefits, and obligations of this instrument shall continue in full force and effect for a period of fifty (50) years. Should the conveyed property be used for any purpose other than provided for in this Amended Supplement to Deed prior to the end of said fifty year period, the conveyed property shall immediately revert and be conveyed in fee simple to the Department of Game of the State of Washington.

It had been our intention from the start that some nonpolitical group should act as operator of the park, and this intention was spelled out in an agreement with the park board in conjunction with the deed. This was in accord with the best practice of most of the world's great zoological facilities, and usually a zoological society is the operating agency. Accordingly, I named the Tacoma Zoological Society in the basic agreement with the Metropolitan Park Board as Connie and I had been active in that society for many years—I as a member of the Animal Committee and twice society president, and Connie as secretary and enthusiastic participant in its education and fund-raising programs. This operating agreement was based on the San Diego model and had been approved by the Tacoma city attorney prior to its submission. It was accepted as evidenced by the signing of the documents by myself and the park board president. But when it came to implementing this agreement it was an entirely different matter. Although several variations were worked out laboriously and later signed, the attorney general's office later reported that the agreement even as amended was not legal in Washington State. This same office had already reviewed the original documents containing the conditions of the gift. We gave up, and it was in fact just as well that we did since the society itself proved too inexperienced and insufficiently funded to carry out the inherent responsibilities, and I conceded that a board of directors, with strong Tacoma Zoological Society participation, would satisfy the terms of the agreement. For a time this board did carry out most of the functions of an operating body.

This was all academic, however, when it became evident that the park board had made no decisions about how to finance the park and planned to carry out only a two-year holding action while they studied the matter. They asked my opinion about the cost of maintaining the status quo, and I suggested that if Charlie McKasson continued as foreman, with my help, we could get along on $25,000 a year, which would cover Charlie's salary and what winter feed and supplies were required. They endorsed this proposal, borrowed the money from a bank, and left Charlie and me pretty much to ourselves for the next two years so far as animal and land management were concerned.

Planning for the future of the park got under way before the Tacoma Metropolitan Park Board determined how to finance the development of the project. Lengthy meetings were held by the members of the Mammal Committee of the Tacoma Zoological Society, and in August 1973, Charles Faust, architect and planner for the San Diego Zoo, joined our study group, providing us with his experience, perspective, and wisdom. He was delighted with the beauty and suitability of the land for use as a wildlife park and felt that the free-roaming concept was not only practical but the only acceptable way of displaying American ungulates. He recognized that, as in all wildlife parks, there would be problems of aggression and veterinary management due to the impossibility of constant observation of the animals.

Encouraged, we continued to plan. Not only were those of us who were involved in the project enthusiastic, but other members of the community became intrigued as well. Frank Jackson, Director of the Washington State University Cooperative Extension Service and originator of the name Northwest Trek, arranged for frequent National Guard helicopter expeditions to the lake, bringing city councilmen, mayors, and an occasional congressman. It was always an adventure as the whirlybirds settled on one of the meadows, flattening the tall grass and blowing seed heads wildly about. Instead of frightening the animals, the elk in particular gathered around the machines so that a guard would always have to stand by to be sure that their curiosity did not lead to damage from the massive antlers of the bulls. The news media were equally enthusiastic about the project. They reported the arrival of new animals on prime-time programs and wrote up favorable stories on events at Northwest Trek. Our national television coverage was launched with a twenty-minute Bill Burrud "Animal World" program, which is still occasionally aired.

Meanwhile, the park board finally decided to finance the development of Northwest Trek by floating a 1.5 million dollar general obligation bond issue. This proposition appeared on the ballot in February 1973 and passed with a resounding 70 percent "Yes" vote. Unfortunately, the necessary 60 percent turnout required for validation was not achieved. Despite this contretemps we were encouraged by the favorable reception and hired planners to start on the complicated task of establishing priorities and designing the park, pending resubmission of the proposition at a later date.

The Richardson Associates, Architects were chosen. They in turn brought Jones and Jones, Environmental Planners, into the picture. As a first step, they undertook a new and far more realistic market and financial study, predicting a break-even revenue by the third year with an attendance of 283,000 persons. This reduced expectation reflected a less optimistic national economy and the beginning of the energy squeeze, as well as the inflation of building costs. Even this prediction proved overly optimistic as far as the date for break-even is concerned. The time seems to be approaching, however, for the park is already 70 percent self-supporting.

In November 1973 the bond issue was resubmitted, for at that time there were other matters of wide interest also on the ballot, increasing the likelihood of validation. I felt strongly that with the rising cost of construction, now estimated to be at least 2.5 million dollars, the public was entitled to a detailed description with graphic representations of what they were being asked to vote for. We again took the plunge and commissioned the planners to start the master plan and design concept immediately, to be completed before the election. Since, as in the case of the county government, no planning funds were available in the park board's budget, two individuals guaranteed a bank loan (handling it through the Tacoma Zoological Society) to provide the necessary funds, which would be repaid if the bond issue were approved but would be considered contributions if the issue failed.

While planning proceeded, many citizens joined in the effort to inform the public of the issues and create enthusiasm. Bill Elder, a retired air force general, directed a magnificent campaign, while Frank Jackson contributed his expertise at every stage. The Tacoma Zoological Society participated and nearly every civic organization, fraternal order, political and educational institution endorsed the project. I was able to work actively on the campaign as well as on-site with the surveyors and planners, helping to lay out the roads and parking areas, marking trees to be removed and, with Charlie, building five and one-half miles of nature trails.

The bond issue passed with 61.2 percent of the vote in November 1973. The planning money was recovered and the deadline for the completion of the park scheduled for the summer of 1975. A furious pace was now set with planning meetings in the afternoons and evenings. Discussions of all aspects of the project were held involving the planners, engineers, architects, Tacoma Zoological Society, and park board representatives, reaching a climax with the presentation of the results of all our labors to the Metropolitan Park District commissioners March 11, 1974. With their approval, ground was broken and immediately thereafter the project got under way.

Working within the animal compound presented some difficulties. That fall the rutting bull moose resented all intruders, and I spent much time protecting the terrified surveyors with a nine-foot fiber glass fishing rod. The whip-

Badger. (Photograph by W. H. Shuman III.)

ping movements and whishing sound are an excellent means of controlling aggression. On the whole, however, road building and animal activities impinged little on each other.

Pine marten. (Photograph by Gary Oberbillig.)

Outside the free-roaming enclosure, the visitor center—with its impressive curved two-hundred-foot-long post and timbered entryway flanked by rest rooms, gift shop, snack bar, and administration offices—began to take shape. The construction of pools and grottoes in the woodland and wetland exhibits came next.

Here in a curving, concrete, tunnellike building, half above and half below ground, a collection of smaller animals indigenous to the Pacific Northwest is housed. Of particular interest is the collection of the mustelid family—the weasels—a fascinating group occupying nearly every niche in the ecosystem: the nearly subterranean badger, the aquatic river otter, the semiaquatic mink, the arborial marten, the slow-moving terrestrial skunk with their unique defense mechanisms, the more generalized fisher and wolverine, and the tiny frenetic weasel that relentlessly pursues its prey in holes underground, in fields and forests, and in the trees. All are fierce predators, with the exception perhaps of skunks and otters. All have beautiful fur, which becomes them far more than it does a human wearer, and most are secretive, rare, and almost impossible to study in the wild. Here in spacious and natural settings, the visitors make their first contact with this legendary family. One can also see beaver in a cutaway of the lodge. They go about their business, grooming themselves and one another with the combed toenail of each hind foot, shredding bark, or swimming above and below water in their large pool. And, of course, there are the raccoons, their clever black hands exploring, their bright eyes inquiring. Adaptable and ubiquitous, they are programmed for success in any changing world.

Between these exhibits and the lake, the land falls off abruptly then levels out in a timbered flat through which a stream flows. This area was enclosed to form the domain of our wolf packs, and when introduced they immediately made themselves at home. They dug their own elaborate dens, where they pupped on two successive years. At night or when distant fire sirens sound, the howling chorus of nine

The dedication of Northwest Trek in 1975.

wolves thrills Connie and me, for it is indeed a ululation of wild eerie harmonics, rising, falling, no two voices on the same frequency, then fading out as suddenly as it starts. The wolves seem content in this home of theirs. They do not pace the fence line as canids do in zoos. They feed on roadkilled deer, which promotes the natural activity of hiding, burying, carrying about, pupfeeding, and maintaining the social hierarchy. Wolves seem to be almost a cult animal, and to judge by the time the visitors spend observing them, one realizes that there is some atavistic bond between modern urban man and the vanishing wolf.

The five and one-half-mile tour route passes along the edge of the lake below these wolf compounds. An articulated propane tram transports visitors on an hour-long tour of the free-roaming area of the park. Here bison, elk, mountain sheep, goats, moose, pronghorns, woodland caribou, and deer move freely, intermingling or isolating themselves as they choose. It is the public that is confined, and the animals that roam free.

Northwest Trek was dedicated on July 17, 1975. Governor Dan Evans did the honors. Representatives of most of the political entities were present on the podium. Speeches were made, we congratulated one another on our foresight, and all in all it was a great day, especially for us and for the many others who had in various ways helped bring it about.

Wind River Range

Little did Connie and I suspect that the official opening of the park would not be the culmination of our efforts but rather the beginning of a five-year personal ordeal. We watched, for the most part helplessly, the perilous struggles of the new park until responsible, competent management could be found and viable working relationships developed with the city of Tacoma and with the Metropolitan Park Board. During this period we witnessed the neglect of the land, inept animal management, and the resignations of many initially enthusiastic and dedicated young naturalists. These five years were like the Dark Ages, filled with murky undercurrents, lacking in clarity of purpose, and tarnished by petty jealousy, power plays, and cover-ups. It was difficult at times to be hopeful about the future.

There were open confrontations at the park board meetings and rumors of similar disagreements within the Northwest Trek board in addition to conflicts between the administration and employees at the park itself. These highly vocal critics and criers of doom were few in number but formed a coalition, which had in some cases overlapping representations on the park board, the Northwest Trek board of directors, and/or the Tacoma Zoological Society and were firmly supportive of the inept administrations of Point Defiance Zoo and Northwest Trek. The press, sensing that there were the makings of a big story and knowing who the protagonists were, set out to investigate with enthusiasm. Since stories of doom are far more attractive to the reader than calmly balanced expositions, the critics had a field day. They predicted decimation of the environment by overgrazing, challenged the free-roaming concept, and foretold of disaster from interspecies aggression. Innuendos, distortions of the facts, and even intemperate personal attacks added spice to the banquet. Editorials and letters to the editor appeared and with each unfavorable story park attendance predictably fell.

The League of Women Voters, keenly aware of the financial and administrative problems at Northwest Trek, undertook an evaluation and finally submitted a report. They asserted that Northwest Trek was an important regional asset that should be preserved but would be administered better at the state level. The league emphasized that the Tacoma Zoological Society's influence should be reduced and offered its help in any way possible. This report helped still the clamor and finally earned recognition from all but the most antagonistic critics. It recognized that the park was indeed a regional asset, and that means must be found to support and develop its potential.

While all this was transpiring, Connie and I often walked with anger and frustration as our companions. I sometimes stood hesitating at the entrance to the long high-walled cul-de-sac called obsession, pulling back only when I glimpsed, fleetingly, the contorted faces that peopled that crepuscular alley, recognizing their nature, and knowing them by name as paranoia. However, we remained convinced that our original concept was valid, and with the anxious support of our daughters together with the loyalty of our friends, colleagues, former patients, and often total strangers, we kept afloat. We found escape when possible in travel to distant places, and I found solace as well in the mountains and wide horizons of Wyoming.

After the park opening and the final agreement within the Northwest Trek board and by the park commissioners that there should be a separate director for Northwest Trek, a search was undertaken. The response was not great. The number of applicants having any real expertise in wildlife management was smaller still, and most of these lacked administrative experience. Finally we chose a candidate whose background seemed closest to our needs. He was hired to start in December 1975.

The interim months were disorganized; most decisions were postponed although some construction work was done. We set up consistent animal care and quarantine procedures and worked out feeding and land management protocols. For a time the new director experienced a smooth transition. Then gradually matters subtly deteriorated. Unionization of part of the employees caused discontent among others. With the establishment of work classifications the feeling of being a family in which everyone lent a hand to whatever project needed doing vanished. Communication began to break down between an increasingly secretive administration and office staff, where nepotism flourished, and the rest of the employees who felt like outsiders. Staff meetings were held rarely. Information was distorted through hearsay and rumor.

All the blame cannot be laid at the door of the administrator, for it soon became evident that although the public enthusiasm for the park was nearly universal, the attendance because of gas rationing at one point, the worsening economy, and many other factors over which the park had no control, fell far short of expectations. Hence revenues failed to cover expenses. This is turn caused irritation and frustration in the park board, already beset by public criticism of the erratic conduct of its affairs. Pressure for a tightened budget increased. Once priorities were established, the animal acquisition and care portion of the Northwest Trek budget suffered most. Its share fell to only about 4 percent of the total—an impossibly low figure in a park where animals were the chief attraction. It was at this point that the greatest internal strains began to show. I pointed out that it was poor economy to try to save on the quality of feed, hay, fertilization of pastures, and acquisition of suitable animals to fill out our collection. Two successive education directors resigned during this time, feeling that they were not able to carry out their missions, and some of the naturalists joined in criticizing the administration and its priorities.

It was evident that Northwest Trek was deteriorating rapidly. Constructive criticism was resented. In the park board, those drawing attention to animal management problems were labeled troublemakers. And little attempt was made to get at the truth. Thus a definite division, becoming at times a confrontation, emerged—in the Northwest Trek board, among the park personnel, and involving myself as well, which prolonged the time when it was recognized that a change was necessary. Conditions became so difficult for dissenters that most of them resigned. I was not reappointed to the Northwest Trek board which increased my isolation from all activities within the park, and made both Connie and me feel strangers in our own country. So we sat in the middle of a fast deteriorating paradise with our dreams turning into recurrent nightmares. It was a bitter time.

Troubles within the park district were not, however, confined to Northwest Trek. The Point Defiance Zoo had for some time been under severe criticism for poor management, erratic animal care, and employee relations. These allegations had been ignored in the same manner as at Northwest Trek by the park commissioners.

This growing disorder could not last. Tacoma's City Council had a great stake in the efficient management of park board affairs since it provided a large portion of the annual park board budget while unhappily having no control over its spending. As arguments waxed louder and more extravagant and the press jumped in to expose, describe, interview, and quote, it became evident that some definitive action was necessary. There were calls for dissolution of the park board with the city taking over its duties. There was a move in the state legislature to dissolve the park board, and even the board itself was split over the need for its existence.

When the tower began to topple, it crumbled fast. Under continuing pressure, the directors of both Northwest Trek and the Point Defiance Zoo resigned. During a stormy session of the park board, the superintendent of parks and the chief financial officer also resigned and were followed in that course by other long-term personnel. Unable to operate without a staff, the park board allowed the city to step in.

It was during this dark and difficult period in Northwest Trek's development when Connie's and my hopes for the park seemed to be crumbling, that I felt a growing paranoia and anger so nearly overwhelming that I knew I had to seek one of my sanctuaries—to find peace and perspective.

I believe that most of us hold in the mind's eye and memory certain intimate places whether they be a particular room, a secret hiding place behind the hay bales in a barn, a shack, or a piece of the outdoors. These places have served as refuges from a world too much with us, as quiet or secret dream spheres where we may pretend everything is as we wish it to be—places beyond the critical eye of family or peers. In my life there have been and still are a good many such precious havens, and almost all of them have been far from the works of man.

My special places have characteristics in common, although they are widely separated geographically and have played important roles over the span of many years. The ground is dry but not parched, carpeted with small plants of many varieties. The grass grows just tall enough to hide a reclining boy or man— for one cannot get close to the earth when sitting in a chair or on a rock or log. Three sides of these small fragrant meadows must be sheltered, whether by boulders, thick brush growing on a bank, but usually by a forest edge. The foreground should slope away gently, but with small shrubs or rocks to define the limits of the immediate domain while not obstructing the wider view. The more distant prospects may vary greatly so long as they remain in a natural state—a slow-running brook, a small alpine lake, a valley, or a range of hills will do. Awe-inspiring mountain peaks, the endless curve of the ocean's horizon, the crashing and pounding of waves on rocky shores, the roar of torrent, are not appropriate backdrops to my special places. Sounds must be harmonious and muted like that of a gentle breeze, the hum of nectar-seeking insects, or the chuckling and murmuring of slow-moving water.

One introduces such places only to special friends or a lover—and then anxiously and defensively like an author first exposing a precious work. Silent appreciation and awareness are the appropriate responses. Except for the rare appreciative ones, the best companions on one's visits to such havens are a dog or a horse. Wordlessly the dog will explore the immediate warm, grassy site and, finding it good, look briefly at the farther scene, then turn slowly three times in a counterclockwise circle and bed down with a contented sigh, his back against a rock or bush. The horse, with ears pricked, also looks out to the larger view, then drops his head and starts to clip and munch the forbs and grass, acknowledging the safety and peace of the surroundings.

I remember one such place at my California school—a grassy little promontory that jutted out from the mesa top not far from the playing fields. Three sides were ringed with sage and mesquite while the fourth was open to the view stretching beyond the sudden drop-off of the mesa rim to the valley and oak-clad hills of the opposite canyon wall. Here I kept two hives of bees. Down the steep mesa wall a hidden little trail led to the half-stone shack I have previously described.

One night when I was thirteen, I awoke sometime in the dark hours, on fire with fever and obviously irrational. One need overpowered my foggy mind—to get to that little promontory and lie in the tall grass: then everything would be all right. Somehow I got out of bed, dressed in a shirt, jeans, and boots and found my way to the barn at the bottom of the hill. With only a rope halter I managed to climb via the corral fence onto my horse's back and head her up the steep trail to the mesa's top and out to my sanctuary. The next thing I remember was being carried in Dr. Manning's arms, and waking up desperately thirsty in a Santa Barbara hospital with a very sore stomach. I was told that my horse had returned to the barn. When I did not appear and my bed was found empty, a search was started, but it took several hours to find me curled up in the tall grass. I had acute appendicitis, and it was only luck that one of my friends reported my special affinity for the promontory.

Many other special places stand out in my mind, although less significantly, like beads on an almost endless string: the tepee site in the Canadian Rockies by the little stream facing Sawback Lake with its backdrop of rocks and snowfields, the many camps beneath sycamores and oaks in the coastal range of California, a few dry desert camps, camps in the Cascades and the Olympics, and in the last fourteen years, camps in the Wyoming Wind River Mountains. It was to that special area in Wyoming that I returned.

My nephew Rob and his wife Martha welcomed me with quiet understanding, providing me with a horse (Leon, an old companion of many previous pack trips), and gave me the key to their cabin at Burnt Ranch. It was on this site along the Oregon Trail that weary emigrants crossed the Sweetwater River for the ninth and last time before heading out into the dry lands on the second half of their journey to the Willamette Valley or branched off on the Mormon route to Salt Lake. History lives at Burnt Ranch. At one time a station was built at the site and used by the Pony Express, the telegraph, and the stage lines. By 1862 a detachment of the Eleventh Ohio Volunteers was stationed there to protect it from Indian attack. When the soldiers left six years later the station was set afire, hence the name Burnt Ranch. All that

Turn-of-the-century Burnt Ranch cabin is situated on the Oregon Trail in Wyoming.

remains is the old cabin, built before the turn of the century, the horse shed and corral, and the deep ruts of the wagon wheels that turned so long ago.

I spent six days alone there, breathing the cleansing air of that high basin country. At first I did little but walk the river, watching the pronghorns that gathered close on the rounded hills behind the cabin. I found a magpie fledgling in the willows and brought it back where it sat on a shelf by the woodstove, reducing the strips of steak and bacon I taught it to eat into massive, smelly excrement. By the second day I felt controlled enough to saddle up and ride out, fording the river where the wagon wheel ruts came to the water's edge.

I wondered at the space, the cloudless sky, the tiny pellets of earth and sand kicked up by my horse's large chipped unshod hoofs. I caught a few horned toads and put them in my shirt pocket where they dozed contentedly. I saw the flash of a tawny body, weasellike with black-tipped tail (but far too large for a long-tailed weasel) bounding through the brush and disappearing into a prairie dog hole. I was suddenly excited, for this must have been a black-footed ferret, rarest of all American mammals, but when I came home to the park, I mentioned it only casually, doubting my own observation. Since that time, black-footed ferrets have been observed and a few captured for study and breeding experiments in an area not far from the ranch. I now dare claim a sighting.

During the next days I rode out farther into the desert and approached the herds of wild horses scattered over the wide expanse of the sagebrush basin. The watchful stallions were always first to spot me and warily watch my approach, keen to discover whether my horse was a mare and thus a potential addition to the harem, or a rival stallion. But when close enough to detect a rider, their ears back and neck curved—snakelike—and they drove their mares and colts in a cloud of dust to a safer distance. Small bunches of antelope were nearly always visible, sometimes quite close and at other times so far away that they could only be identified when the sun touched their white rump patches. Sage hens burst in a booming flurry of wings from close underfoot. Rabbits, ground squirrels, and prairie dogs watched my approach, then, with a flip of tail, ducked into their mounded burrows.

As the soft breeze swept across my face, tempering shimmering heat, I felt a composure flow slowly through my body and my mind like a benison. The soft, shuffling sound of unshod hoofs, the smell of horse sweat and bruised sagebrush were my close companions. The far expanse of flatland bounded by the distant outline of the Oregon Buttes ahead and the jagged snow-capped Wind River Range to the east beckoned, offering new adventures for another time. There was a flash of light far out toward the buttes and through my binoculars I recognized the typical shape of a sheepherder's wagon from whose small window the sun was reflected. A white horse was picketed nearby.

When I returned to the cabin, the small special area of

meadowland stretching to the thick, willow-bordered river, I unsaddled, turned my horse into the old corral, and felt almost at peace. As I approached the building, I noticed a slight movement at the corner of the house. A tail-wagging, tentative, and utterly unkempt Australian sheep dog met me. She could only have come from that distant sheep camp, having perhaps spotted the light of my Coleman lamp during the night across the miles of crystal clear high-country air.

Of course I knew that the loss of a sheep dog to a herder who, at most, probably worked with only two or three, was a serious matter. Although I was delighted to have the companionship of the dog, I made no great effort to encourage the little bitch. The next morning I saddled up, and with the dog at heel, started the long cross-country ride toward the sheep camp. In about an hour I could see plainly the wagon with its rounded top and brightly painted stripe of red to provide easy identification at a distance. Soon the constant bleating of hundreds of sheep could be heard; the little dog began to quiver and whine. Then, irresistibly drawn, she streaked off in the direction of the sound and her inborn destiny.

As I came up to the wagon the herder appeared, riding his second horse. He waved cheerfully, beckoned me, and while dismounting, he half scolded and half praised his wayward dog, obviously delighted to have her back. He told me he was Spanish, but he may have been Basque as were so many sheepherders. He thanked me for bringing back the dog, as indeed he had been hard put to control his flock with only one old dog and an impetuous young pup. The sheep had scattered and flowed and billowed toward a low depression in the land where a shallow water basin shrank from its fissured mud-edged banks.

Dogs are as essential to a shepherd as horses to a cattleman in open range country, but whereas the small operator

Joe, a sheepherder along the Wind River Range in Wyoming, peers from his immaculate little wagon.

of a well-fenced and well-organized cattle spread can get along fairly well without a horse, a sheepman needs his dog under nearly all conditions. Many ranchers feel, and I agree, that herding dogs, no matter how well trained, are a handicap when working cattle, especially when there are calves in the herd. In certain circumstances they may be useful to drive cattle out of steep draws or dense brush, but cows are made aggressive and nervous by dogs that may spook the herd, and a swift kick can easily damage the most agile cow dog.

Sheep are a different matter. Although a little border collie may weigh only fifteen to twenty-five pounds, a fraction of the weight of his charges, he can outwit, outbluff, outrun and, if necessary, punish even the largest and most unruly sheep in the flock. There are many breeds of sheep dogs, somes specializing in different aspects of sheep handling, but I refer here to the smaller types—the border collie and some of his variants such as the Australian sheep dog. I have seen these work in England, Wales, Scotland, the American West, Australia, and New Zealand. Everywhere they show the same nearly supernatural skills at controlling sheep. You can spot the puppies with outstanding potential when still very young for they have, as one New Zealand trainer at a government sheep experiment station explained it, "the eye." I had been keenly aware of "the eye" without defining it in my mind, for we had owned a gentle little border collie, Maggie, who had it to a degree.

"The eye" is the expression of an instinct to face down and force one's will on another creature by fixing it with a steady gaze of such burning intensity that it cannot be resisted. Usually this stare is delivered from a crouch, belly to ground, perhaps the rest of the body quivering slightly like a tightly wound spring but held in check by the chock of the fixed eyes. Even the angriest old ewe, stamping her foot and threatening to charge, will face this confrontation only briefly before turning aside. Small puppies with the deep, inbred herding instinct and "the eye" try their skills on any moving object. They will even outstare chickens and ducks. Only beetles and butterflies seem to ignore the mysterious force that they project.

I have often watched Maggie when she had nothing else to do dominating my horse, Annie, while she grazed quietly in the horse pasture. Maggie would approach her in a mock stalk, belly close to the ground, then, when about six feet away, flatten out and give the horse the full blast of her stare. Annie apparently paid no attention for a minute or two, except for a slightly accelerated cropping, then as though unconcerned, turn aside and graze slowly in the opposite direction. Maggie then would rise quickly, move to a new position, drop and stare, and thus drive the horse from one end of the pasture to the other. No sound was made, no sign of impatience shown by horse or dog, but there was never a doubt that the little dog had total control over the eleven-hundred-pound horse. When herding, sheep dogs instinctively work to bring the sheep toward the shepherd, and so when training young dogs especially, he keeps to the front or side of the flock, capitalizing on this trait. I sometimes think that the state department should send its bright young prospects to sheepherding school and select for important posts only those who have "the eye."

Joe, my newfound friend, invited me into the immaculate interior of his little wagon. It was of standard design, mounted on automobile front wheels and axles with shafts for horse pulling and a trailer hitch for longer moves when towed by a pickup. Across the front a built-in bunk served also as a bench with storage beneath. There were cupboards on the walls, and a small cast-iron stove with chimney protruding through the roof to provide heat and cooking facilities. A small hinged table and shelves, a woodbox, and a powerful and modern radio completed the furnishings. He said that he listened mostly to broadcasts from Red China. He did not understand them, but he liked the music and the voices. Many of his remarks were difficult for me to comprehend, as his accent and syntax combined to tantalize without clarifying, but we drank coffee and talked as best we could. He had a lot to say about "gomens," which turned out to mean "women," and we parted, mutually rewarded by the contact, yet each ready to resume the simpler companionship of his own company.

The next morning I awoke to the squealing of horses and the hollow sound of hoofbeats on packed earth. Rob and Martha's stallion, a fine bright chestnut quarterhorse, had brought his little band of mares and colts and was trying to pick a fight through the high pole fence with my saddle horse. I had trouble driving him off, but I did not want to find myself in the middle of a fight when I rode out later in the day.

That evening I carried my little tame magpie down to the river's edge, found the huge, complex nest from which he had strayed and near which three other fledglings sat in a bush, waiting for their parents' return. There I released him. I also freed the horned toads, bulging with the flies and other insects I had found for them. Before I fell asleep that night a coyote called from the rocky hill behind the cabin where the crumbling piles of stones mark the graves of unnamed Oregon Trail travelers. Its call bounced from ridge to ridge and multiplied into a wild chorus of yips and yodels. I slept dreamlessly, and the next morning Rob's pickup appeared over the rise. We jumped Leon into the truck bed between the flimsy stock racks and headed back to Lander and the ranch. I was whole again.

Northwest Trek: An Enduring Reality

IN THE SPRING OF 1982, three and one-half years after my Wyoming retreat, Connie and I were returning from a trip to England. As our plane made its gradual descent, we watched the Cascade Range grow before us. The volcanic peaks dressed in wispy clouds appeared magical and unreal. We approached the steep asymmetrical summit of Mount Hood, then Mount Saint Helens, her familiar, near-perfect cone blasted into a truncated cauldron by its eruption. Mount Adams, partly obscured by clouds, appeared serene by comparison.

Soon Mount Rainier was at eye level with the plane. Its massive flanks and glorious triple crown occupied the entire view on the starboard side. Snow extended almost to its base and the glaciers sparkled in the sunlight that filtered through the clouds. This great volcanic upthrust of the earth is the grandest mountain in the lower forty-eight states. Foothills rise from the flatlands of the north and west in soft undulations, rapidly becoming more angular, rugged, timbered, until they merge to form the mountain's base. Small blue lakes dot the hollows between, while silvery twisting ribbons of rivers all run westward to the ocean or the inland sea. Straight-edged patches of clear-cut forest are like pieces of a jigsaw puzzle from the air—brownish yellow where newly cut, purple-pink where alders infiltrate the newly planted or spontaneously seeded Douglas fir. Darker green areas of genetically vigorous and uniform second growth contrast with the more ragged and rapidly vanishing old growth.

Our descent became more abrupt, the rolling map beneath us more detailed, and suddenly, almost directly below and fewer than twenty miles from the base of the mountain as the bald eagle flies, our eyes caught the familiar configuration of the tiny section of land, which has been our lodestone and intermittent home since 1937.

Only our intimacy with its every contour could make this small microcosm of land stand out. There indeed lay the pie-shaped plateau with two irregular areas of intense green pastureland that we had cleared with such labor in the fifties and early sixties. And between them rose the steep, flatiron-shaped ridge crowned by a pale growth of red-barked madrona trees. The small, blue horseshoe-shaped lake was clearly visible. Looking closer we recognized the swamps, bogs, beaver ponds, and pothole by the foliage surrounding them. Though obscured by fir trees, our house could be distinguished in the north indentation of Horseshoe Lake. In the large meadow that overlooks the steep valley of Ohop Lake, we saw tiny brown dots, too dark for elk, and therefore the park's herd of bison. A few seconds later we were over the flatland of the Puyallup Valley, bright with patches of yellow daffodils and red tulips. On the port side there was Tacoma. Point Defiance, green with old growth, enclosed Commencement Bay on one side. The logged abutments of Brown's Point and Dash Point bounded the other. Beyond lay the straits, inlets, and watery arms of Puget Sound.

I had made this descent many times, picking out the familiar landmarks as if they were coordinates on a map. It helped me to get my bearings. But this time something was different. My recognition was keener, my perception sharper. It was as if distance and time away suddenly helped me to see anew this land so familiar and beloved and yet the focus of so much anxiety and frustration for the past six years. The plane landed, and soon we were home again on Horseshoe Lake. Drawn to the window overlooking the lake, I sensed that same heightened recognition. There, once again, the subtlety and beauty of the lake came alive. The light looked brighter, colors sparkled, and the air felt crisper. It welcomed me in a way that I had not felt for a long time.

Still wearing my travel clothes, I climbed into my sand-colored Jeepster Commando—my park vehicle—and set off. I wanted to reassure myself that spring was coming, the grass was growing, and the animals had wintered well. The park was closed to the public during the week at that season, so I had the tour route to myself. I could pretend that all this lovely place was mine again—mine and the animals' with no need to share. Tomorrow I would return it without regret to the people to whom we had given it and who come to Northwest Trek—some joyfully for a family outing, some to increase their understanding of the world of nature, and still others to maintain a link with the past when wilderness existed close at hand and was, as it must remain, a natural

Horseshoe Lake, Northwest Trek Wildlife Park, 1985.
(Photograph by Tirrell and Richard Kimball.)

and national heritage. But not now—not today.

As I joined the paved, one-way, five and one-half-mile tour route, I could smell the sweet balsam of Peru odor from the sticky buds of the black cottonwood trees. Along the edge of the lake the alders showed pinkish green swellings where the leaves were ready to unfurl, their catkins hung like yellow tassles, a few dark woody cones remaining from the previous year. Clumps of green Indian plum—the first signs of spring—showed small clusters of white flowers between their leaves.

As I drove slowly, not twenty feet from the lake's edge, it was obvious that the waterfowl were at the height of their mating activities. Pairs of Canada geese searched for nesting sites, the long-necked ganders "wonking" their high-pitched note, while the smaller, more compact geese consulted constantly with their mates. One gander stood on top of a huckleberry-covered stump and picked up pieces of rotted wood, while his goose, who apparently found this selection undesirable, called him in low-pitched tones.

Mallard drakes, resplendent in breeding plumage, swam about on the lake in groups or in solitary bachelorhood, often chasing each other and doing battle; but when a hen appeared, great excitement resulted. Heads began to bob to the side, and there were frantic individual or gang attempts at mating.

I watched four trumpeter swans carve smooth furrows through the glassy surface of the water. They sculled with one black foot, the other tucked along the side, and uttered their loud cornetlike call. They too, felt the stirrings of spring; in this southern latitude, formal courtship and breeding has occurred, and on occasion, successful nesting.

A great blue heron stalked in the shallows beside the beaver lodge, his neck half-extended, the gorgeous mating plumage on his crest and breast revealed. There is no more skillful or intense stalker than the great blue heron. I turned off the ignition to watch the perfection of his hunting. His yellow, unblinking eyes are set close beside a rapier beak that

Trumpeter swans. (Photograph by W. H. Shuman III.)

forms the armed tip of a projectile, spring-loaded by the curving neck, which is anchored between strong shoulders. As he moved almost imperceptibly forward, his long toes drew together scarcely rippling the water and, emerging again, they only dimpled the surface with each new stride. This heron is an old friend that spends most of the year along the lakeshore and has learned to ignore our close approach.

It was time to move on, and as I passed the first little swamp on my right I noticed that a goose was already setting, her head tucked along her back and eyes closed. The gander stood tall at the end of the mossy log nearby.

Then crossing the grassy knoll, I entered the twenty-acre meadow. This evening the bison herd was scattered across its lush turf of orchard grass and clover. The herd bull was lying down, a monolithic dark hump, tufts of winter hair already dull, shedding from his back. The cows and younger bulls were all grazing, and characteristically, ignored the jeep. The flanks of the cows were full, foretelling a good calf crop in May. What a stirring sight it always is to watch these shaggy beasts, unchanged from earliest times, afraid of nothing, dangerous but dull-witted, again reclaiming isolated portions of their ancestral lands after near extinction less than a century ago.

The road continued through a deep, shady patch of mixed woodland that forms a forest edge to the meadow. I almost missed seeing the bull moose, standing half-hidden in the shadows, his dark body seemingly detached from the ground by the lighter colored legs. This creates effective camouflage in deep woods where often only the flick of a long, mulelike ear or other motion catches the observer's eye. He, too, seemed to have wintered well, his antlers thick and symmetrical, his pendulous "bell" still weighted with fat. Emerging again from the woods, the road skirted the steep, open face of the ridge where four California bighorn sheep were nibbling new green shoots. The three-year-old ram had a good half curl to his horns, and two of the three ewes appeared pregnant.

Rounding the toe of the ridge, the shaded wetlands provide an abrupt contrast. The first of these wetlands are deep black pools surrounded by ferns, and at this time completely hemmed in by the large, exotic-looking leaves of skunk cabbages, their bright yellow flowers the very symbols of spring. Next the road skirted the long peat bog, which is divided partway by a timbered peninsula, and crisscrossed by large rotting logs covered with salal. Pond weeds, bulrushes, water lily, and duckweed offer food and shelter for a great variety of insects, amphibians, and invertebrates here. The waterfowl, muskrat, beaver, as well as white-tailed deer and moose also enjoy the varied aquatic vegetation, while the several old snags still standing offer homesites for the tree-nesting ducks. (In fact, at that moment, I heard the wistful call of a wood duck alarmed by the approaching jeep.)

The road then doubled back along the far side of the bog between two more wetland areas. The first is rain fed, home to a pair of hooded mergansers. The second is the beaver pond formed about twenty years ago when two year-old beaver kits from the big lodge on the lake were driven out, as is the way of beaver families, and had to find new lodgings. Over the years they have built a solid dam, raising the water to about four feet, drowning out the alders. Slowly the cattails and marsh grasses have come in, and more recently water lilies. I noticed that the beaver lodge had vegetation growing on it and so is not inhabited at this time, for active lodges are kept well plastered with mud and free of grass and weeds. A gander anxiously patrolled the far shore of the beaver pond, but I did not spot his setting mate and continued around the end of the pond.

The road was built over what was a lower secondary beaver dam. When it had been a jeep road, a hollow cedar log acted as a culvert so that the water would not overflow and wash out the road. But every night in spring the beaver plugged up the log and every morning we unplugged it. And now, with a good blacktop surface and a steel culvert, the beavers still plug it every night and we unplug it every morning. We need not and cannot win, but we must stay even to keep the tram route passable. The water here is at the very edge of the road, which, during the summer months, allows visitors in the trams to see turtles, bullfrogs, western newts, muskrat, and occasionally mink. This close viewing is always particularly exciting for schoolchildren.

In the second series of meadows beyond the ponds, the elk herd, including four bulls, their antlers tumescent under the soft velvet, lay in the new grass. Cows and calves were with them, chewing contentedly on their cuds. I looked up at the threatening skies, because this simultaneous lying down often predicts imminent rain.

At the forest's edge behind the elk a family group of white-tailed deer ignored the jeep. The only other wildlife I could see were violet-green swallows exploring woodpecker holes in an old snag, and beyond, a greater sandhill crane challenged me, throwing back his scarlet-capped head with each raucous cry. Beside the road, a black-tailed doe with her nearly grown twin fawns was sampling the new runners of trailing blackberry, and they paid little attention to the

Black bear (photograph by Dick Milligan.)

passing jeep.

The road climbs, traversing the south face of the ridge and skirting the upper side of the grove of red-barked madrona trees. This grove lends an exotic quality to the surrounding forest, for its varicolored green and red bark, contorted trunk and limbs, and shiny leaves appear tropical in contrast to the vertical and spiky shapes of most northern trees. In March the madronas are covered with clusters of white blossoms, which are a favorite food of band-tailed pigeons. The fragrant blossoms that escape the pigeons turn into clusters of bright orange berries enjoyed by most species of birds including the flickers and woodpeckers. Even the berries that fall to the ground and ferment are not wasted, for robins in particular feast on these, becoming harmlessly intoxicated in this free "happy hour." Thus for nearly three months of the year, the madrona is a major food provider for the birds at Northwest Trek.

I reached the end of the ridge where it breaks out into the open allowing a dramatic two hundred and fifty-degree

Black-tailed deer fawn. (Photograph by Gary Oberbillig.)

view, and looked directly ahead at the rounded Bald Hills extending in gradually lighter shades of green, fold upon fold, to the horizon. To the left in the foreground, the steep south face of the ridge plunged two hundred feet to the plateau of the meadow, where the bison were still grazing. Beyond the meadow, again the land dropped abruptly three hundred feet to the Ohop Valley below, where the park property ends at the county road.

Rising steeply from the far side of the valley, a two thousand-foot range of foothills blocks more distant views with its undulating outline. The silhouette of this hill has always reminded me of the torso of a recumbent woman. I have thought that some Puyallup or Nisqually Indian camping by our little lake some long-ago summer must, too, have recognized this outline and called it "Sleeping Woman." Perhaps he and his family felt kinship with this supine woman shape. Like us, perhaps they found in the little lake and it surroundings a place where bush, tree, tiny plants, and the animals from the smallest shrew to the great bull elk are all sheltered in the woman's shadow. It seemed to me, at that moment, that the earth breathes, and only the boulders around me sleep.

Twilight comes fast in the dark forest. I was reminded of this, for I heard the long mysterious quavering note of a varied thrush. This sound is not like that of any other bird, for it has a ventriloquistic quality always seeming to come from a great distance. It is uttered only in spring and chiefly early in the morning and again at sundown from the topmost limb of a dominant tree. This thrush is a shy bird and its "song"—a prolonged note followed by a second pitched an interval lower—though evocative, does not display the versatility of other members of the thrush family. Perhaps its bright orange breast, black collar, and varied plumage render vocal accomplishments superfluous in the mating game. As if in answer, a winter wren fluttering not three feet from the ground beside the jeep, burst into urgent song—scolding, tinkling, burbling—for more than six seconds. Like a brown fluttering troglodyte it disappeared into a clump of shiny salal leaves.

I was nearly home, warmed, refreshed, and reassured by what I had experienced. The animals had wintered well and would thrive. Northwest Trek Wildlife Park would thrive as well. It was an enduring reality. If beauty is in the eye of the beholder, my "new view" of Northwest Trek had found it. We had, I realized, lost our perception of it during the years when problems and uncertainties occupied our lives and dulled our vision. But now all the doubts and frustrations seemed less relevant. It was indeed a renaissance.

Since the administrative changes in 1982, there has been relief on all levels. The political maneuverings and the tugs of war have subsided. The few strident voices of dissent have stilled, at least for a time. The new management has been enlightened and competent. And at last, there is a sophisticated understanding of the necessity of reconciling the ever-changing environment with the requirements of sound wildlife management. But above all, part of our dream has come true. Northwest Trek has earned recognition as a unique and successful experiment in new techniques for displaying American wildlife in its natural setting, while, at the same time, providing unparalleled opportunities to increase man's understanding of his relationship with the natural world and celebrate the wilderness experience that is our natural and national heritage. Northwest Trek has become the perfect setting for studying animal behavior and physiology; research not possible in the wild. For three successive years, a federal agency, the Institute of Museum Services (I.M.S.), has awarded Northwest Trek a $75,000 grant to expand and enrich the park's on site and outreach educational programs. This grant has already affected thousands of Northwest school children. It also utilizes local college and university cooperative education interns and professional staff.

As a pediatrician-naturalist, the educational aspect of the park is of particular interest to me, and I enjoy involvement in it. The quality of the material, both from the standpoint of art and exposition, delights me. It is only the beginning of what can be developed in the way of graduate level research in wildlife management and soil, forest, and wetland ecology—areas which are being explored with concerned agencies at both the state and federal level. We expect their involvement to be continuous.

Recently Northwest Trek, after the long and arduous accreditation process, has received the highest recognition from the American Association of Zoological Parks and Aquariums, endorsing the concept and praising the facility and its management—to me a long awaited vindication.

People of this region as well as visitors from around the world express delight in the beauty of the terrain and the wildlife that moves freely within its boundaries. And Connie and I derive great pleasure from the realization that thousands of children, escaping the confinement of the schoolroom to visit the park, get their first intimate experience of wildness, while the aging and socially isolated come and recall with shining eyes a past embroidered by time and take Northwest Trek to their hearts.

Connie and I remain for our lifetimes in the lovely house we built on the small brown lake, observing and recording the activities of the wildlife all around us while it largely ignores our presence. I am privileged to contribute to the planning and philosophy of the park as its honorary resident naturalist. Retired from active medical practice and teaching, announcements of births; bar mitzvahs; graduations; and weddings continue to evoke warm and colorful memories of my pediatric years. We are more free to pursue our other interests in travel, friends, music, books and, of course, our children and grandchildren, all of whom feel a sense of pride and involvement in the park.

But the lake and the surrounding land remain our niche where we find our greatest fulfillment and reward as only happens when man, his works, and nature are enmeshed in the common web of life.

David and Connie Hellyer at Northwest Trek in the 1990s.

Lightning Source UK Ltd.
Milton Keynes UK
UKOW01n0609130117
292002UK00008B/139/P

9 780295 996493